Springer Climate

Series Editor

John Dodson ⓘ, Institute of Earth Environment, Chinese Academy of Sciences, Xian, Shaanxi, China

Springer Climate is an interdisciplinary book series dedicated to climate research. This includes climatology, climate change impacts, climate change management, climate change policy, regional climate studies, climate monitoring and modeling, palaeoclimatology etc. The series publishes high quality research for scientists, researchers, students and policy makers. An author/editor questionnaire, instructions for authors and a book proposal form can be obtained from the Publishing Editor. **Now indexed in Scopus® !**

More information about this series at http://link.springer.com/bookseries/11741

John C. Shideler · Jean Hetzel

Introduction to Climate Change Management

Transitioning to a Low-Carbon Economy

 Springer

John C. Shideler
Futurepast Inc.
Arlington, VA, USA

Jean Hetzel
Johanson International
Paris, France

ISSN 2352-0698 ISSN 2352-0701 (electronic)
Springer Climate
ISBN 978-3-030-87920-4 ISBN 978-3-030-87918-1 (eBook)
https://doi.org/10.1007/978-3-030-87918-1

This Springer imprint is published by the registered company Springer Nature Switzerland AG
The registered company address is: Gewerbestrasse 11, 6330 Cham, Switzerland

Earth provides enough to satisfy every man's needs, but not enough for every man's greed.
Mahatma Gandhi

In 2020 the objective of limiting warming to less than 2 °C by 2100 compared to the pre-industrial era is no longer realistic; we are on a path at the global level to 3.5–4 °C.
Jean Jouzel
June 5, 2018, in Paris

Entire systems are collapsing, we are at the beginning of a mass extinction, and all you can talk about is money and fairytales of eternal economic growth, how dare you?
Greta Thunberg
United Nations, New York, September 2019

Preface

This book updates and rewrites the authors' recent book *Le grand livre de la stratégie bas carbone: Principes et outils,* published in France in April 2019. How much difference two years make! The SARS-CoV-2 virus was first reported in China in late 2019 and during the first quarter of 2020 spread around the world. The World Health Organization declared it a global pandemic in March 2020. Governments confined citizens to their residences and skies cleared of pollution as travel by air, rail, and road ground to a halt. "Nonessential" businesses closed to protect workers from contracting COVID-19, the disease caused by the novel coronavirus which proved to be highly contagious and deadly.

In the first and second quarters of 2020, economies around the world shrank in proportions not seen since the Great Depression of the 1930s. Business activity shriveled in sectors such as travel and leisure, restaurants and hotels, and personal cosmetic services. The oil and gas industry was hard hit, with prices for West Texas Intermediate crude oil dropping into negative territory in April 2020 because storage capacity was at or near limits. Some parties with future contracts for May delivery paid buyers to take delivery of oil on their behalf. White-collar employees worked from home, while hospital and other "essential" workers braved exposure to the virus and were disproportionately sickened. By mid-2021, several million people around the world had died from COVID-19, and many more millions had become unemployed.

While the economic depression caused by COVID-19 devastated lives and even more livelihoods, it also temporarily "bent the curve" on rising emissions of carbon dioxide, methane, and other gases that are chiefly responsible for climate change. The pandemic of 2020 jolted people in the world's privileged countries into awareness of the fragility of economic prosperity. The unprecedented speed with which vaccines against COVID-19 were developed and authorized for use encouraged many to ask when our economies would "return to normal" and how long that would take. The authors of this book choose to ask a different question: "How can political and business leaders in a post-pandemic world reconstruct economies that both deliver economic security in an equitable way and address the urgent climate crisis that poses an existential threat to our future?"

The French edition of this book was shaped in part by discussions the authors participated in at the end of 2017 when the French government organized a "Climate Finance Day" in Paris to consider how a niche but growing interest in "green" financial instruments could be encouraged. The authors are involved in helping both to write international standards on green debt instruments including green bonds and green loans, and to define a "taxonomy" of sectors in which investments of capital can contribute to the decarbonization of economic activity. According to climate scientists, transitioning to a low-carbon economy in the next three decades is the only way that Earth's warming can be limited to 1.5–2 °C by 2050. At present rates of warming, the world's population is like frogs placed in a warm pot of water with the heat gradually rising. The frogs may not feel an immediate need to jump out of the pot, but it is clear that their fate is cooking to death if they remain inside it.

Our ambition for this book is to suggest that the post-pandemic rebuilding of the world's economies can do better than "return to normal." As the COVID-19 pandemic ravaged economies in the second quarter of 2020, Christine Figueres, former executive director of the UN Framework Convention on Climate Change Secretariat, wrote the following: "The next ten years will determine whether we stand any chance of preventing the worst impacts of climate change, orders of magnitude worse than the COVID-19 disruption. If by 2030 we have not cut greenhouse gas emissions by half globally, we will not be able to avoid devastating tipping points that would shatter the global economy and pose existential human threats. The costs of inaction are staggering—$600 trillion by the end of the century."[1] We agree with Ms. Figueres that there is no time to waste in addressing the world's climate crisis, and that the world's economic recovery from the COVID-19 pandemic provides an impetus to redouble efforts to reform the world economy in ways that are both environmentally sound and socially responsible. We believe that such a future is possible if there is political will to achieve it.

We are indebted to the work of countless climate scientists, policy thought leaders, economists, journalists, and others whose work we cite in this book or whose knowledge and wisdom guided our intellectual development over the years. To all who labor in the interest of making the world a better place to inhabit for humans, animals, and plants, we extend our gratitude. We also acknowledge the help of two associates of John Shideler who assisted in preparing the final manuscript: Thank you Kate Puddy and Callie Pople. We are solely responsible for any errors that may remain .

Arlington, VA, USA   John C. Shideler
Paris, France   Jean Hetzel

[1] Christine Figueres, "Choices made now will shape the global economy for decades to come," Financial Times, 2020-05-08, p. 17.

Contents

About the Authors

John C. Shideler, Ph.D. is President of Futurepast Inc., a consulting firm located in Arlington, Virginia. He is an environmental professional with a diverse background in sustainable aviation fuels, green finance, management systems, and auditing. He focuses on issues related to combatting climate change and the transition of economies to a sustainable future. He serves as a lead validator and verifier of greenhouse gas projects, as a verifier of statements related to green debt instruments and climate actions of financial institutions, and as a management system consultant and auditor. He has helped write numerous international standards relating to climate change and environmental performance and serves as Chair of ISO's subcommittee 4 on environmental performance evaluation. He is a consultant to policy makers in the aviation sector and to national accreditation bodies. He is Author of several works of history and serves as Adjunct Professor in the College of Engineering Technology at Rochester Institute of Technology where he teaches courses on greenhouse gas management. He may be contacted at john.shideler@futurepast.com.

Jean Hetzel is Chairman of JOHANSON International, a Paris-based consulting firm, and an environmental expert with finance and legal training. He has evaluated the financial consequences of water pollution at the French state ministry level. A pioneer in environmental auditing in Europe, he performed engagements for the French environmental program of the Centre National de la Recherche Scientifique and as an international auditor for many major industrial companies. He is also a building and construction expert with experience performing integrated building system certification audits. After five years of technical research, he worked on the design of many green buildings for several companies and communities. One of his buildings was certified as HQE 14/14 high level (equal to LEED Platinum). He has been a lead contributor to the development of the ISO 14030 Part 3 taxonomy, forthcoming. He also serves as a lead verifier for green bond and green loan engagements. He has served as Associate Professor (Cergy and CNAM) and Advisor for large companies in Europe and some large cities. He is Author of several technical books in French. He is affiliated with Finance Watch, a European Finance NGO. He is an accomplished sailor. He may be contacted at j.hetzel@johanson-international.com.

Abbreviations and Symbols

$	United States dollar
€	Euro
AR4	Fourth Assessment Report (of the IPCC)
AR5	Fifth Assessment Report (of the IPCC)
BCE	Before the Common Era
C	Celsius
C$	Canadian dollar
CFC	Chlorofluorocarbon
CO_2	Carbon dioxide
COP	Conference of the Parties (to the UNFCC)
ECP	Extended concentration pathway
GHG	Greenhouse gas
GWP	Global warming potential
HCFC	Hydrochlorofluorocarbon
HFC	Hydrofluorocarbon
IAM	Integrated assessment model
IPCC	Intergovernmental Panel on Climate Change
Mt	Million tons
NDC	Nationally determined contribution
RCP	Representative concentration pathway
SCEP	Sustained adoption of soil carbon enhancing practices
t	Metric ton
UNEP	United Nations Environment Program
UNFCCC	United Nations Framework Convention on Climate Change
WMO	World Meteorological Organization

Note on Language and Numbers

This book was written in American English, which is native to one of the co-authors. Accordingly, we use the word "billion" to refer to 1000 millions, and "trillion" to refer to 100,000 millions. We use commas to separate thousands and periods to indicate the start of decimal places in numbers.

Chapter 1
The Science Background

1.1 Climate Change and Global Warming

The issue of climate change has increasingly occupied the attention of both scientists and policy makers since the last quarter of the twentieth century. The point in time when societies reached a "tipping point" for demanding action on climate change mitigation will be determined by future historians. A likely candidate will be the years between the negotiations of the Paris Agreement in 2015 and 2021 when activist youth, commitments of business leaders, and increasing manifestations of the effects of climate change raised societal awareness about the urgency of taking action. According to science historian Thomas Kuhn, holders of established beliefs and value systems will ignore scientific evidence that challenge their convictions until the evidence becomes so overwhelming that a paradigm shift can no longer be avoided (Kuhn 1996). The world's population and their governmental representatives should by now be well past that tipping point as the protests of climate skeptics become fewer and less convincing.

1.1.1 The Work of Climate Scientists

As early as the 1820s, French scientist Joseph Fourier realized that certain atmospheric gases shrouded the planet like a transparent bell jar admitting sunlight and absorbing infrared rays. He understood that the atmosphere was heated from both above and below: first, by sunlight as it shone through the jar, and second, by the infrared rays that the Earth radiated back into space as it cooled overnight.

In the 1860s, Irish scientist John Tyndall conducted experiments to measure the amount of radiant heat (infrared radiation) that certain gases could absorb and transmit. He found that water vapor and carbon dioxide were good absorbers and emitters of infrared radiation (NASA 2016).

© The Author(s), under exclusive license to Springer Nature Switzerland AG 2021
J. C. Shideler and J. Hetzel, *Introduction to Climate Change Management*,
Springer Climate, https://doi.org/10.1007/978-3-030-87918-1_1

Later in the nineteenth century, Svante Arrhenius, a chemist who became Sweden's first Nobel prizewinner, was destined to have a bigger impact than he possibly could have imagined. Building on the work of John Tyndall, Arrhenius associated the temperature of the planet to the amount of CO_2 emitted into the atmosphere (Barral 2019). He is arguably the father of climate science.

Knowledge of climate science continued to grow in the twentieth century. In the mid-twentieth century, Charles David Keeling began measuring carbon dioxide in the atmosphere at the Mauna Loa Observatory in Hawaii. Keeling, a researcher in the Scripps Institution of Oceanography at the University of California, San Diego, used an infrared spectrophotometer for continuous measurement. He discovered that CO_2 emissions were higher at night when plants and soil released CO_2 via respiration. The values observed in the afternoon represented "free atmosphere" CO_2 concentrations over the Northern Hemisphere. Keeling's measurements were influential because they showed steady rates of increase of atmospheric concentrations of CO_2 over several decades.

1.1.2   Atmospheric Concentrations of CO_2

The mechanisms for maintaining the Earth's moderate temperatures are now well understood. The Earth is warmed by solar radiation which reaches the planet's surface through atmospheric "windows" at varying wavelengths in a spectrum that ranges from very short (X-rays) to very long (radio waves). About 70% of the sun's energy reaches the Earth's surface to warm it. Approximately 20% is reflected back into space as infrared radiation. This radiated energy from the Earth's surface passes through our atmosphere where some of the energy is absorbed by greenhouse gases. As greenhouse gas emissions have risen to historically abnormal levels, the atmosphere—and thus the greenhouse gas effect—has increased to levels not previously experienced in Earth's geological history (Whetstone 2021; NASA 2016).

Historical data demonstrate the extent to which concentrations of CO_2 in the atmosphere have exceeded previous levels. In Fig. 1.1, the graph on the left shows evidence that during the last 800,000 years of the Pleistocene era, atmospheric concentrations of carbon dioxide fluctuated in a range from about 175 parts per million by volume (ppmv) to 300 ppmv until the twentieth century when concentrations accelerated rapidly. According to the US National Aeronautics and Space Administration (NASA), CO_2 concentrations reached 412 ppmv in 2019, a 47% increase since the beginning of the industrial age (Buis 2019). As a point of reference, our human species, *Homo sapiens*, evolved about 200,000 years ago or about three-quarters of the way from left to right in the left-hand graph in Fig. 1.1.

The right-hand graph in Fig. 1.1 shows the steady rise of CO_2 emissions since 1950. The rise from approximately 300 ppmv to nearly 420 in 2021 is particularly significant because CO_2 has a residence time in the atmosphere of between 300 and 1000 years. According to NASA, "Half of the increase in atmospheric carbon dioxide concentrations in the last 300 years has occurred since 1980, and one quarter

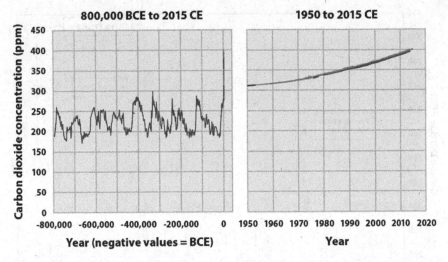

Fig. 1.1 Historical data over millennia (before the common era) and during a recent 65-year period (US EPA 2016a)

of it since 2000" (Buis 2019). The long residence time of CO_2 in the atmosphere means that reducing emissions in the twenty-first century—even to zero—will not immediately end the warming effect of CO_2 that humans have already emitted to the atmosphere.

The problem of heat-trapping greenhouse gas emissions is not limited to carbon dioxide. Other gases also contribute to global warming. Among them is methane, which traps more heat than CO_2 during a much shorter residence time in the atmosphere. Methane is better known as "natural gas," the fuel commonly used to power electricity generating stations and to warm residential homes, among other uses. "Methane concentrations have increased 2.5 times since the start of the Industrial Age, with almost all of that occurring since 1980. So, changes are coming faster, and they're becoming more significant" (Buis 2019).

One sees in Fig. 1.2 that atmospheric CO_2 concentrations over the last 400,000 years never exceeded 300 ppmv until modern industrial times. Since then, they have increased by a third and now exceed 400 ppm. By April 2021, average concentration levels had reached 419 ppm (CO_2 Earth 2021) with *Homo sapiens* responsible for the additional 25% of concentration beyond the historical maximum (Freedman and Mooney 2020a). On the timescale of Fig. 1.2, the recent increases in atmospheric CO_2 appear as a straight vertical line.

The publication in 1998 of a graphic representation of variations in temperature over the thirty-year mid-twentieth-century average (1961–1990) made plain to everyone how unusual the rise in global temperatures had been since the beginning of the Industrial Revolution. The charted temperatures evoked a "hockey stick." Cooler temperatures from the Middle Ages to early modern times resembled the handle, and

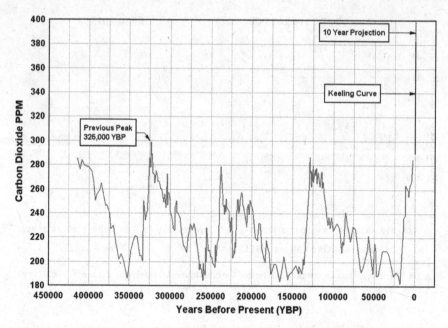

Fig. 1.2 Atmospheric carbon dioxide during the last four ice ages (Planet for Life 2017)

the chart of temperatures rising abruptly upward since 1850 looked like the blade (Mann et al 1998) (Fig. 1.3).

The "Keeling Curve" in Fig. 1.4 records the steady rise in concentration of atmospheric CO_2 since the late 1950s (Planet for Life 2017). The Keeling Curve shows a cyclic variation of about 5 ppmv each year corresponding to the seasonal change in uptake of CO_2 by land vegetation in the Northern Hemisphere. From a maximum in May, the level decreases during the northern spring and summer as new plant growth takes CO_2 out of the atmosphere through photosynthesis. After reaching a minimum in September, the level rises again in the northern fall and winter as plants and leaves die off and decay, releasing CO_2 back into the atmosphere (Climate Central 2013). This fluctuation explained the Keeling Curve's "sawtooth" pattern of emissions concentration (Fig. 1.3).

Charles Keeling monitored data at the Mauna Loa volcano in Hawaii. Figure 1.4 clearly indicates a small—but detectable—impact on CO_2 concentration of Mount Pinatubo's volcanic activities. After lying dormant for almost 500 years, the volcano in the Philippines erupted in June 1991. In one of the most destructive volcanic eruptions of the twentieth century, thick deposits of tephra (rock fragments and particles ejected by a volcanic eruption), streams of pyroclastic flow, and lahars (mudslides) caused significant damage to the economy and infrastructure of surrounding cities. The volcano also ejected millions of tons of sulfur dioxide into the atmosphere,

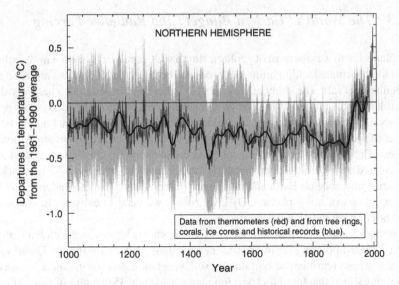

Fig. 1.3 CO_2 concentrations and average temperatures since the year 1000 (Mann 2016)

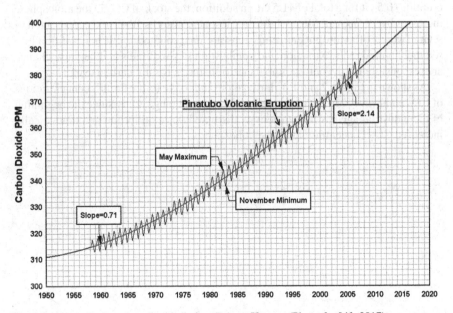

Fig. 1.4 Atmospheric carbon dioxide during the past 50 years (Planet for Life 2017)

which contributed to ozone depletion and caused a worldwide decrease in temperature (Mount Pinatubo 2011). The graph also shows that the rate of increase in atmospheric CO_2 concentrations has risen in the years since data collection began.

1.1.3 The World's "Carbon Budget" and Radiative Forcing

Our planet Earth was for most of geologic time a well-regulated system whose carbon emissions fluctuated within limits that produced both tropical conditions and ice ages at regular intervals. The study of ice cores in glaciers and in Antarctica has provided valuable insights into this variation. With carbon concentrations in 2020 at 417 ppm of carbon dioxide in the atmosphere, scientists express confidence that humans have exceeded previous high points during known geologic time (Freedman and Mooney 2020b). The extent of human influence has caused some to conclude that we have now entered a new geologic era, that of the "Anthropocene," or a geologic age marked by human influence on Earth and its climate (Carrington 2016). To understand why concentrations of atmospheric CO_2 keep rising, we need to consider how human activities perturb the global "carbon cycle."

Figure 1.5 shows the annual fluxes of CO_2 in what is known as the global carbon cycle. Upward pointing arrows represent emissions to the atmosphere. Downward pointing arrows represent carbon sinks in soil, vegetation, and oceans and waterways. This figure shows that fossil fuel development and use add 35 gigatons (Gt—or billion tons) of CO_2 per year to the atmosphere, land use change adds 6 Gt, and volcanic eruptions 0.5 Gt for a total of 41.5 Gt. In addition, the stock of CO_2 in the atmosphere increases by 18 Gt/year. Meanwhile, the biosphere absorbs 12 Gt and oceans another 9 Gt. Two Gt of anthropogenic emissions are unaccounted for. It is not difficult to see that humankind's extraction and combustion of fossil fuels and land use change are responsible for the current imbalance of nature's natural carbon cycle.

As shown in Figs. 1.2 and 1.3, during thousands of years the Earth's concentrations of CO_2 emissions varied within a relatively constant range. Only since the beginning of the industrial age has human activity disrupted this equilibrium and dramatically increased CO_2 concentrations in the atmosphere.

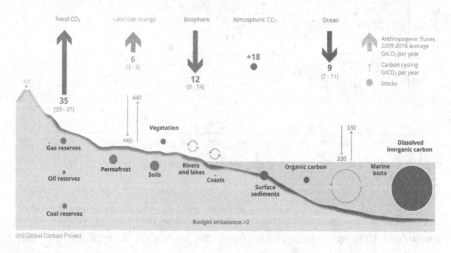

Fig. 1.5 Anthropogenic GHG emissions and sequestration during 2009–2018, GtCO$_2$/year (Global Carbon Project 2020)

Describing the residence time in the atmosphere of CO_2 is not straightforward. Approximately 70% of CO_2 molecules emitted to the atmosphere are reabsorbed within 100 years of emission. Another 10% are reabsorbed within 300 years. The remaining 20% reside in the atmosphere for tens or hundreds of thousands of years (Hausfather 2010). Given the role that marginal emissions of CO_2 have in increasing atmospheric CO_2 concentrations, it is understandable that scientists would assign the primary responsibility for global warming to releases of CO_2 from fossil fuel use and deforestation.

Other gases also contribute to Earth's greenhouse effect. Methane (CH_4) and nitrous oxide (N_2O) also absorb heat radiated from the Earth's surface and contribute to global warming. Methane is not only extracted from gas-bearing geologic formations as a fossil fuel. It also is produced from the decomposition of organic material where it is sometimes known as "swamp gas." Animal husbandry, especially the raising of cattle, is an important source of methane emissions along with rice cultivation and municipal solid waste landfills. Less well known are methane clathrates that exist in colder regions of the ocean such as the Arctic and on land in permafrost. Clathrates are crystalline ice structures that contain molecules of other chemicals, in this case methane. Estimates of the global mass of methane clathrates range from 0.5 to 10 trillion metric tons. Cold temperatures and pressure maintain their integrity, but global warming increases the risk of their future release into the atmosphere (Lambert et al. 2006 pp 12–13) if actions are not taken to limit global warming.

Methane is one of several gases in a class of short-lived radiative forcers (SLRFs). Unlike carbon dioxide, methane molecules reside in the atmosphere for about 12 years. Over a twenty-year time horizon, they are 86 times more powerful than carbon dioxide molecules. For this reason, it is particularly urgent to mitigate their emissions in the period from 2020 to 2050 when the world's efforts are focused on reducing GHG emissions in the amounts necessary to limit global warming to no more than 2 °C.

Nitrous oxide is emitted from plants as they fix nitrogen fertilizers in their roots as well as during certain industrial processes such as nitric acid production. N_2O resides in the atmosphere for about 109 years before it is reduced to chemically smaller units via the process known as photolysis (Prather et al. 2015).

Because different GHGs reside in the atmosphere for varying lengths of time and absorb heat radiated from the Earth at different rates, the Intergovernmental Panel on Climate Change (IPCC) has normalized these characteristics to make them comparable. Since the IPCC's Second Assessment Report (SAR) in 1996, the IPCC has published "global warming potentials" (GWPs) for greenhouse gases (IPCC 1996). Subsequent assessment reports have recalculated global warming potentials based on more recent science.

Global warming potentials describe how much energy a greenhouse gas absorbs in the atmosphere over a specific time horizon, relative to CO_2 (US EPA 2020). Radiative forcing is directly related to heating of the Earth's surface, as it measures the extent to which more solar energy is absorbed by Earth than is reflected back into space. Radiative forcing measures the heating effect at planet Earth's surface caused by transfers of energy from greenhouse gases to other gases in the atmosphere,

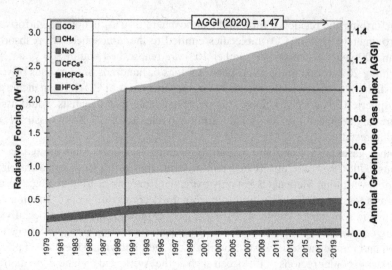

Fig. 1.6 Radiative forcing caused by major long-lived greenhouse gases (NOAA 2021a)

primarily nitrogen (Whetstone 2021). Figure 1.6 shows the amount of radiative forcing caused by various greenhouse gases, based on the change in concentrations of these gases in the Earth's atmosphere since 1979.

The data in Fig. 1.6 come from a variety of historical ice core studies and recent air monitoring sites around the world (US EPA 2021a).

Radiative forcing is calculated in watts per square meter (W/m^2). It measures the amount of thermal radiation reflected from the Earth's warmed surface that is absorbed by greenhouse gases in the atmosphere. On the right side of the graph in Fig. 1.6, radiative forcing has been converted to the Annual Greenhouse Gas Index, which is set to a value of 1.0 for 1990 (NOAA 2016).

The Earth's atmosphere is comprised primarily of nitrogen (78%) and oxygen (21%), with other gases making up the balance. Carbon dioxide is present in the atmosphere at concentrations of only about 0.04%, yet it has an important effect at the margins in warming the atmosphere and indirectly the Earth. Our planet is inhabitable precisely because the Earth's atmosphere traps a majority of the sun's incoming radiation. Without this "greenhouse" effect, life on Earth would be much more difficult to sustain.

Some thermal radiation reflected from the Earth's surface passes through the atmosphere into space, but a portion of it is absorbed by atmospheric greenhouse gases. When infrared photons from the Earth's surface collide with undisturbed CO_2 molecules in the atmosphere, energy is transferred to the molecule increasing its vibrational and rotational energy. The now excited CO_2 molecule transfers its energy to the much more abundant nitrogen and oxygen molecules. As the temperature of a gas is a function of its average velocity (speed), molecules of nitrogen and oxygen become warmer (Whetstone 2021). Figure 1.8 helps to visualize the process (Fig. 1.7).

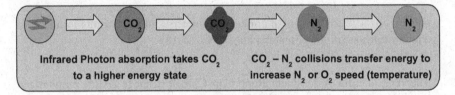

Fig. 1.7 Process for energizing a CO_2 molecule and transferring its energy to a nitrogen molecule (Whetstone 2021)

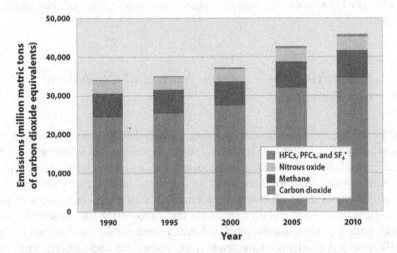

Fig. 1.8 Global greenhouse gas emissions by gas, 1990–2010 (US EPA 2014)

Radiative Forcing Concept

A radiative forcing is an energy imbalance imposed on the climate system either externally or by human activities (e.g., changes in solar energy output, volcanic emissions, deliberate land modification, anthropogenic emissions of greenhouse gases, aerosols, and their precursors). A radiative forcing is something that can usually not be observed but provides a simple quantitative basis for comparing the response in global mean temperature to different imposed agents (Sorteberg 2020).

The term "radiative forcing" has been employed in the IPCC Assessments to denote an externally imposed perturbation in the radiative energy budget of the Earth's climate system. Such a perturbation can be brought about by secular changes in the concentrations of radiatively active species (e.g., CO_2 and aerosols), changes in the solar irradiance incident upon the planet, or other changes that affect the radiative energy absorbed by the surface (e.g., changes in surface reflection properties). This imbalance in the radiation budget has the

potential to lead to changes in climate parameters and thus results in a new
equilibrium state of the climate system (IPCC 2018).

The IPCC's Fifth Assessment Report on Climate Change (AR5) emphasized
how radiative forcing influences the Earth's surface temperatures. The aspirational
numbers referred to in the Paris Agreement—global warming not to exceed 2 °C and
preferably to rise no higher than 1.5 °C—are based on measured global mean surface
temperatures. Localized temperature increases or decreases represent variations from
the mean.

1.1.4   Quantifying Anthropogenic Sources of Greenhouse Gas Emissions

Global warming measured in watts per square meter describes the problem caused
by anthropogenic emissions of GHGs. It is not, however, a useful indicator for GHG-
emitting organizations that wish to take actions to address the problem. More actionable are the actual GHG emissions that organizations make directly through their
own activities and those of their energy suppliers, customers, and supply chains.
For this reason, GHG inventories are based upon accounting for emissions whose
global warming potentials are normalized over a standardized time horizon (generally 100 years). A shorter time horizon may be appropriate where an organization is
primarily concerned about methane emissions which have a much shorter residence
time in the atmosphere (US EPA 2021b).

For management purposes, GWPs published in Assessment Report 4 (2007) and
Assessment Report 5 (2013) are now the most widely used. Table 1.1 shows how
the values published in AR5 produce the following comparative results for CO_2 and
CH_4 over 100- and 20-year time horizons:

As illustrated in Table 1.2, organizations more commonly use a 100-year time
horizon to quantify the impact of greenhouse gas emissions.

Table 1.2 does not provide GWPs for two categories of "engineered gases,"
hydrofluorocarbons (HFCs) and perfluorocarbons (PFCs). HFCs and PFCs define
categories of gases with widely different GWPs, so they do not lend themselves to
brief descriptions. A list provided by the GHG Protocol, which gives GWPs from

Table 1.1 Global warming potentials of CO_2 and CH_4 over different time horizons

100-year residence time		20-year residence time	
GHG	GWP	GHG	GWP
Carbon dioxide (CO_2)	1	Carbon dioxide (CO_2)	1
Methane (CH_4)	28	Methane (CH_4)	86

Table 1.2 Global warming potentials of common GHGs over a 100-year time horizon

GHG global warming potentials averaged over a 100-year residence time per AR5

GHG	GWP	GHG	GWP
Carbon dioxide (CO_2)	1	Hydrofluorocarbons (HFCs)	a
Methane (CH_4)	28	Perfluorocarbons (PFCs)	a
Nitrous oxide (N_2O)	265	Sulfur hexafluoride (SF_6)	23,500

[a] HFCs and PFCs are *categories* of gases, whose many different variants have widely different global warming potentials

the IPCC Second Assessment Report (SAR), the AR4, and the AR5, identifies 19 separate HFC chemical compounds in this category. It also identifies eight different PFCs plus four other gases that belong to the same chemical family. The GWP of HFCs in this list ranged from as low as 4 to as high as 12,400. For PFCs, the range was from 6630 to 23,500 (Greenhouse Gas Protocol 2016).

Figure 1.8 shows worldwide emissions of carbon dioxide, methane, nitrous oxide, and several fluorinated gases from 1990 to 2010. For consistency, emissions are expressed in million metric tons (Mt) of carbon dioxide equivalents (CO_2e). These totals include emissions and sinks due to land use change and forestry (US EPA 2014). Figure 1.8 shows the strong growth in emissions over a 20-year period of the two major GHGs (carbon dioxide and methane) normalized to CO_2e units.

1.1.5 Representative Concentration Pathways

In its recent assessments of climate change, the IPCC has preferred to report on global mean average warming rather than focusing on increased absolute anthropogenic CO_2 emissions. This shift focuses on the effect of global warming rather than on its cause. International negotiations in 2015 resulted in agreement to limit warming to well below 2 °C compared with pre-industrial times. Current IPCC assessment reports now describe representative concentration pathways (RCPs). Scenarios are based upon a time-series projection of emissions and concentrations of the full suite of GHGs and aerosols and chemically active gases, as well as land use/land cover (Moss et al. 2010).

RCPs are created using integrated assessment models (IAMs), and they project outcomes up to 2100 under differing GHG emission scenarios. Extended concentration pathways (ECPs) extend the RCPs from 2100 to 2500 based on stakeholder consultations and professional estimates. Four RCPs from published literature were used in AR5 (IPCC 2014a) as a basis for the climate predictions and projections developed in Working Group I, Chapters 11 to 14:

- RCP2.6 One pathway where radiative forcing peaks at approximately 3 W/m^2 before 2100 and then declines (the corresponding ECP assumes constant emissions after 2100)

– RCP4.5 and RCP6.0, two intermediate stabilization pathways in which radiative forcing is stabilized at approximately 4.5 W/m^2 and 6.0 W/m^2 after 2100 (the corresponding ECP assumes constant concentrations after 2150)
– RCP8.5 One high pathway for which radiative forcing reaches greater than 8.5 W/m^2 by 2100 and continues to rise for some amount of time (the corresponding ECP assumes constant emissions after 2100 and constant concentrations after 2250).

Representative Concentration Pathway

The word representative signifies that each RCP provides only one of many possible scenarios that would lead to the specific radiative forcing characteristics. The term pathway emphasizes that not only the long-term concentration levels are of interest, but also the trajectory taken over time to reach that outcome (Moss et al. 2010).

Figure 1.9 tracks the rise of global CO_2 emissions from 1960 to 2018. It projects CO_2 emissions to 2100 on the basis of three IPCC RCP scenarios:

RCPs provide policy makers with a basis for planning mitigation targets. Risk managers use RCPs to anticipate potential levels of financial losses and to make the necessary provisions to offset them. Investors use RCPs to confirm or inform the potential payback of an investment. RCPs are useful projections, but they do not provide localized information. For example, Arctic Circle regions in Alaska, Canada, and Russia are warming at approximately twice the rate of the "representative" pathway concentrations (Post et al. 2019).

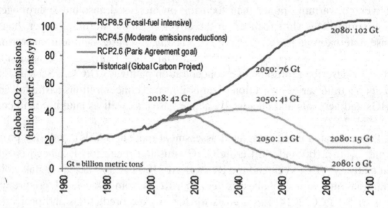

Fig. 1.9 IPCC scenarios (Morse 2018)

1.1.6 Global Warming Hotspots

When scientists and policy makers discuss global warming, the most common approach is to refer to global average temperatures from land and ocean monitoring sites around the world. Thus, targets to limit global warming to 1.5 °C or 2 °C do not refer to absolute limits in every location. Indeed, there are locations around the world where warming has already exceeded these limits. We know, for example, that higher than average warming has occurred in Arctic regions and some locations in the Middle East. Data from September 2020 showing locations with above average warming as well as below average warming are shown in Fig. 1.10.

The computation of global warming on a global average basis is a complicated exercise involving the calculation of local temperature increases and decreases during the course of a year, followed by a consolidation of the local results to a global average. The cumulative year-over-year changes are then compared to a baseline usually represented by data from an average of data from the period 1850 to 1900. The latter 50-year period is used by climate scientists as a proxy for "pre-industrial" conditions before 1750 (IPCC 2014b, p. 50 note 8).

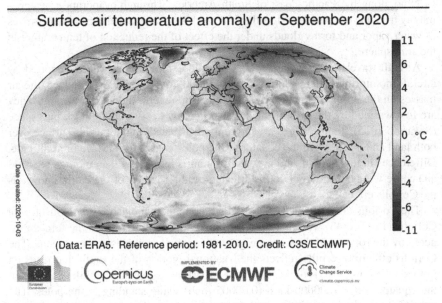

Fig. 1.10 Localized variations from average September 2020 temperatures worldwide (C3S/ECMWF 2020)

1.2   How Climate Change Affects Weather Conditions

The previous paragraphs discuss climate change at the global level. However, global mean surface temperatures mask considerable variation at the local or national level. Figure 1.10 provides a global map of temperature differences from the local average observed in the single month of September 2020. It is notable that the highest temperatures in September relative to the average are located close to the two poles, in northern Siberia and at the South Pole. We will see in the following paragraphs the consequences of these temperature differences on long-term climate equilibrium.

Modern meteorology can be dated to the seventeenth century. The Medici network, founded by Ferdinand II de Medici (1610–1670), Duke of Tuscany and his brother Prince Leopold—both pupils of Galileo (1564–1642)—included 11 measuring stations located in Italy, Austria, Poland, and France (Pratt 2012). Modern meteorological networks are now linked through membership in the World Meteorological Organization (WMO).

Weather mechanisms are well known. Oceans are traversed by cold and warm currents (Kattie 2019). Familiar examples include the Gulf Stream of the West Indies along the USA and Canada that warm the ocean on the Atlantic coast of Europe and El Niño along the Pacific coast of South America. Through evaporation, water— mainly from the oceans (three-fourth of the earth's surface) but also from land—rises as water vapor and forms clouds under the effect of the reduction of temperature in the atmosphere.

Aircraft travelers are aware of the very cold external temperatures reported by aircraft monitoring systems during flights 10,000 m above sea level. Indeed, the air masses on the ground and partially, on the ocean, receive a double warming. They are first warmed by the rays of the sun and then by heat radiating back into space (Savarino 2015). Temperatures are unevenly distributed on the Earth's surface over both land and water, from the equator to the poles, and at the surface or at altitude. Differences in temperature and pressure (hot air is less heavy than more dense cold air) cause winds. Wind direction is affected by the rotation of the earth, known as the Coriolis effect.

The Coriolis effect, named after the French mathematician Gaspard Gustave de Coriolis (1792–1843) who first described the phenomenon, reflects the forces triggered by the rotation of the Earth on gas masses such as clouds and on oceans. The Coriolis effect powerfully reflects the rotational motion of the Earth that leads an individual to move at 1664 km per hour (1040 miles per hour) while standing at the equator and at 0.00008 kmh (0.00005 mph) while standing at the poles. The rotational effect with the influence of depressions (cooling air pressure) and high-pressure areas (high and warming atmospheric pressures) accentuates the phenomena that cause hurricanes. Cold air that is heavier than warm air slips into the space left free by warm air and causes winds, which result from the temperature difference between the pressure radians (NOAA 2021) (Fig. 1.11).

The difference in localized rates of warming from the monthly average level in September (see Fig. 1.10) is only one particularity of global warming. Temperature

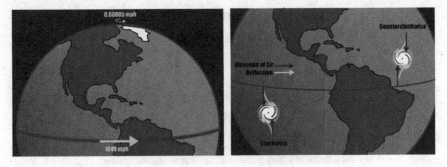

Fig. 1.11 Coriolis effect and influence on hurricanes (NOAA 2021b)

differences of more than 55 °C (131 °F) can be seen from one region of the globe to another. The average value, obtained through global observations, including at the poles, is used to measure increments of global temperature change.

The rise in global mean surface temperature of more than a degree from the pre-industrial era (1880–1900) to the present resulted from an increase in energy in the atmosphere. Energy transfer has occurred based on exchanges between hot and cold masses. The oceans play a key role as an energy regulator by mixing and storing masses of hot and cold water. This function provides acceptable living conditions for humans and other species. Over the last 140 years, the oceans and atmospheric air masses have warmed very gradually. Nonetheless, this resulting temperature increase has affected snowfalls and precipitation rates, and modified habitats and biodiversity. Habitat stress can lead to species extinction, which in turn promotes invasive species that accentuate the imbalances of life (Lindsey and Dahlman 2021).

Warming has had an unwelcome effect in Alaska and Siberia where areas of permafrost are thawing. Thawing of permafrost releases methane trapped for centuries in the soil. The released methane is a fugitive emission of a greenhouse gas that can fuel forest fires in marshes and is more readily ignited at higher temperatures. The methane released has an outsized medium-term effect on climate change because it has a GWP of 86 times CO_2 over 20 years and 28 times CO_2 over 100 years as described in Tables 1.1 and 1.2. Rises in temperatures augment the intensity of wind and precipitation events and increase the frequency of sudden cold snaps.

1.3 The Interdisciplinary Nature of Climate and Earth Science

Climate change is a function of anthropogenic influences on planetary forces that interact with each other and influence the climate that human beings experience on Earth. It is not always sufficiently stressed that the part of the planet we inhabit, a quarter of the earth's surface, is extremely fragile and spatially limited. Scientists refer to this area as Earth's "critical zone".

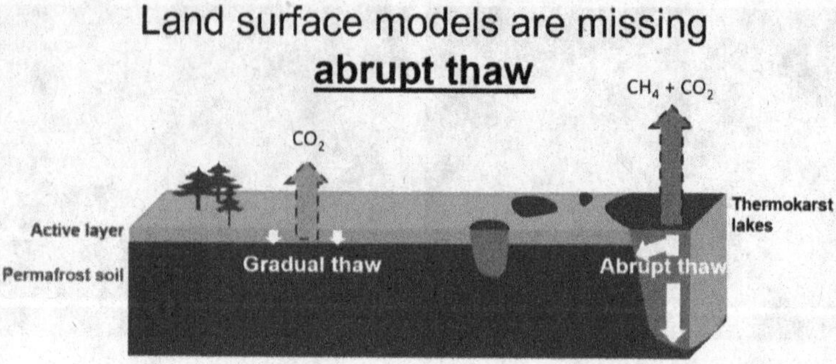

Fig. 1.12 Process that accelerates the release of methane into the atmosphere (Holdren 2021)

The understanding of this concept came to us during an audit we carried out for the French National Center for Scientific Research (CNRS) on its environmental program (Hetzel 1990) under the leadership of Claude Lorius. Lorius is a French glaciologist and inventor of the analysis of air bubbles contained in ice that reveals information about the Earth's climates that existed millennia before modern times.

The interpretation of the effects of climate on the limited layer "between the sky and the rocks"[1] calls upon many scientific disciplines illustrated by Fig. 1.12 (CRITEX 2021). To fully understand the impacts from emitting CO_2 into the atmosphere on this "sky–earth" space requires contributions from the following scientific specialties (CRITEX 2021):

- The science of living things and their environments (ecology) because of the sensitivity of species (trees, plants, insects, animals) to air quality and exchanges resulting from evaporation and photosynthesis.
- Microbiology provides an understanding of the evolutions of the first organic cells created by physical–chemical phenomena and bacteria that, like what happens in the human body, nourish and destroy living plants and organisms to create a chain of life.
- Meteorology explains the thermal and gas flows that cause precipitation through evapotranspiration.
- Hydrometeorology studies the effect of precipitation on soil erosion and on the supply of freshwater for human and animal consumption.
- Pedology analyzes the transfer of CO_2 in the soil and the physical–chemical exchanges resulting from the creation and retention in groundwater as well as the production of methane and carbon in the soil.
- Geophysics studies interactions with deep areas of the earth, in support of biogeochemistry, hydrogeology, and geology.
- Hydrology studies the flow of rainwater and the formation of rivers and lakes.

---

[1] A term referring to part of the Earth inhabited and utilized by humans and other species, from mountain tops to the ground upon which we live, farm, and mine.

The interactions are complex and necessitate a systems view of the environment and climate change. "Systems thinking is about understanding the complex, nonlinear, and interconnected system in which an organization operates" (ISO 2020 p 41). We return to the application of systems thinking in Chap. 7 on adaptation.

1.4 Melting Permafrost in the Arctic

Until recently, permanently frozen land in the higher latitudes of the Northern Hemisphere had survived cycles of cooling and warming over millennia. With global warming, however, permafrost is thawing in places like Alaska and Russian Siberia. A "feedback loop" is becoming established in the warming regions of the far north where thawing of the permafrost leads to emissions of greenhouse gases, leading to more warming, and thus more emissions. The risk is significant, as the amounts not only of methane— representing 40% of the world's reserves—but also carbon dioxide and nitrous oxide could be released, as well as pathogens (Struzik 2020).

Figure 1.12 illustrates the acceleration of methane release during the scenario of abrupt thaw of permafrost that recreates the old wetlands that were caught in the ice. Thermal shocks act upon the subsoil, breaking down its elements. The result is emissions of CH_4 as well as CO_2.

Permafrost thawing accelerates climate change due to increased carbon emissions from wetlands (Fig. 1.13). Permafrost is a methane reservoir that contains a large amount of organic compounds in the icy soil estimated at 1400 GT. Carbon has been buried in permafrost by processes that have occurred over thousands of years. During the last ice age, large ice caps covered most continents. As they spread out and then receded, the heavy ice fields shredded the rock below them into a very fine dust called loess or glacial flour. The ice caps produced an enormous amount of this powdery rock, and the wind and rain blew it to the ground (Schaefer 2021). Permafrost participates in the carbon cycle as soil carbon can be released due to increased surface temperatures under multiple RCP scenarios (see Fig. 1.9).

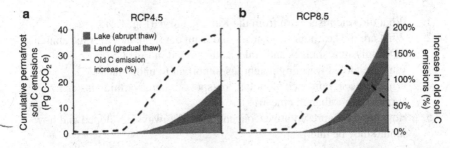

Fig. 1.13 How important is a sudden thaw for permafrost carbon feedback? (Anthony et al. 2018)

1.5   Desertification of the Sahel

The term "desertification" applies not only to the Sahel (southern Sahara) but also to tropical arid areas, identified by a hot and dry climate, mainly in Africa but also in the USA, Australia, and Asia. These areas are characterized by low rainfall, which may persist for several years. They cover 46.2% of the Earth's surface and are home to 3 billion people (IPCC 2019). From 1980 to 2000, arid surfaces have expanded by 9.2%, reducing vegetation productivity. Human factors such as overexploitation of groundwater resources and the growth of invasive plants have weakened biodiversity and the ability to grow subsistence crops. Sandstorms increase soil erosion accentuating the character of "arid land."

There are several reasons why desertification is increasing:

- Deforestation from the use of fire to clear land for agriculture
- Overgrazing
- Inappropriate irrigation
- Intensive crops
- Temperature rises resulting from climate change
- Long-term reduction in precipitation in different regions as a result of changes in ocean climate patterns and the El Niño effect.

To reduce desertification, authorities began a tree planting initiative that was originally intended to create a 15-km-wide transcontinental forest belt from Dakar to Djibouti. This Great Pan-African Green Wall was designed to be as continuous as possible, but breaks could be made if necessary to bypass obstacles (brooks, rocky areas, mountains) or to connect inhabited areas (Escadafal 2012).

We have had the opportunity in this chapter to provide only a brief introduction to the science of climate change. A basic understanding of the science of climate change and its multiple facets is needed to make informed contributions to climate change management—the topic we address in the remainder of this book.

Questions for Readers

1.   What did scientists learn from the Keeling curve (Fig. 1.4)?
2.   What are the key points to understand from data showing average annual CO_2 emissions sources and sinks as illustrated in Fig. 1.5?
3.   Why are global warming potentials important to understand?
4.   What is the significance of global "hotspots" where warming is occurring at rates higher than the mean?
5.   How are the Representative Concentration Pathways developed and how should they be interpreted?

References

Anthony K et al (2018) 21st-century modeled permafrost carbon emissions accelerated by abrupt thaw beneath lakes. https://www.nature.com/articles/s41467-018-05738-9#citeas. Accessed 23 Feb 2021

Barral M (2019) Svante Arrhenius, the man who foresaw climate change. BBVA open mind 19 Feb 2019. https://www.bbvaopenmind.com/en/science/leading-figures/svante-arrhenius-the-man-who-foresaw-climate-change/. Accessed 18 June 2021

Buis A (2019) The atmosphere: getting a handle on carbon dioxide. Accessed 5 May 2020

C3S/ECMWF (2020) Copernicus climate change service. Surface air temperature anomaly. https://climate.copernicus.eu/sites/default/files/2020-10/map_1month_anomaly_Global_ea_2t_202009_v02.pdf. Accessed 11 Oct 2020

Carrington D (2016) The anthropocene epoch: scientists declare dawn of human-influenced age. https://www.theguardian.com/environment/2016/aug/29/declare-anthropocene-epoch-experts-urge-geological-congress-human-impact-earth. Accessed 11 May 2020

Climate Central (2013) Keeling curve. https://www.climatecentral.org/gallery/graphics/keeling_curve. Accessed 5 Nov 2020

CO_2 Earth (2021) Earth's CO_2 https://www.co2.earth/. Accessed 28 May 2021

CRITEX (2021) The critical zone of the earth. https://www.critex.fr/what-is-critex/la-zone-critique-en/. Accessed 15 Feb 2021

Escadafal R et al (2012) The African great green wall project what advice can scientists provide? https://www.researchgate.net/publication/334414300_The_African_Great_Green_Wall_project_what_advice_can_scientists_provide. Accessed 9 March 2021

Freedman A, Mooney C (2020a) Major new climate study rules out less severe global warming scenarios. https://www.washingtonpost.com/weather/2020/07/22/climate-sensitivity-co2/. Accessed 22 Aug 2020

Freedman A, Mooney C (2020b) Earth's carbon dioxide levels hit record high, despite coronavirus-related emissions drop. https://www.washingtonpost.com/weather/2020/06/04/carbon-dioxide-record-2020/. Accessed 28 Aug 2020

Global Carbon Project (2020) Global carbon budget. https://www.globalcarbonproject.org/carbonbudget/20/files/GCP_CarbonBudget_2020.pdf Accessed 18 May 2020

Greenhouse Gas Protocol (2016) Global warming potential values. https://www.ghgprotocol.org/sites/default/files/ghgp/Global-Warming-Potential-Values%20%28Feb%2016%202016%29_1.pdf. Accessed 26 May 2020

Hausfather, Z (2010) Common climate misconceptions: atmospheric carbon dioxide. https://www.yaleclimateconnections.org/2010/12/common-climate-misconceptions-atmospheric-carbon-dioxide/. Accessed 27 June 2020

Hetzel J (1990) Rapport des auditeurs du programme interdisciplinaire de recherche sur l'environnement. (PIREN) CNRS 1990. Unpublished confidential document.

Holdren J, Natali S, Anthony KW (2021) Science session: thawing arctic permafrost—regional and global impacts. https://www.belfercenter.org/publication/thawing-arctic-permafrost-regional-and-global-impacts. Accessed on 24 May 2021

IPCC (2014a) AR5 climate change 2014: impacts, adaptation, and vulnerability. https://www.ipcc.ch/report/ar5/wg2/. vol. 14, 836–840. Accessed 15 Feb 2021

IPCC (1996) Global warming potentials: IPCC second assessment report. https://unfccc.int/process/transparency-and-reporting/greenhouse-gas-data/greenhouse-gas-data-unfccc/global-warming-potentials. Accessed 17 May 2021

IPCC (2014b) Mitigation of climate change. Working group III contribution to the fifth assessment report of the intergovernmental panel on climate change. In: Edenhofer O (ed) https://www.ipcc.ch/site/assets/uploads/2018/02/ipcc_wg3_ar5_full.pdf. Accessed 31 May 2021

IPCC (2018) Summary for policymakers. In: Masson-Delmotte V et al (eds) Global warming of 1.5 °C. An IPCC special report on the impacts of global warming of 1.5 °C above pre-industrial levels and related global greenhouse gas emission pathways, in the context of strengthening the global response to the threat of climate change, sustainable development, and efforts to eradicate poverty. https://www.ipcc.ch/sr15/chapter/spm/. Accessed 15 June 2021

IPCC (2019) Desertification. https://www.ipcc.ch/srccl/chapter/chapter-3/. Accessed 20 Feb 2021

ISO (2020) Guide 84, Guidelines for addressing climate change in standards. https://www.iso.org/standard/72496.html. Accessed 26 May 2021

Kuhn T (1996) The structure of scientific revolutions, 3rd edn. Illinois, Chicago

Lambert G et al (2006) Le méthane et le destin de la terre: les hydrates de méthane: rêve ou cauchemar? EDP Sciences, Paris

Lindsey R, Dahlman LA (2021) Climate change: global temperature. https://www.climate.gov/news-features/understanding-climate/climate-change-global-temperature. Accessed 16 March 2021

Mann M et al (1998) Global-scale temperature patterns and climate forcing over the past six centuries. Nature 392(1998):779–787

Mann M (2016) IPCC third assessment report. https://michaelmann.net/sites/all/themes/themeimg/research_photos/IPCC_2001_3rd%20Assessment_Report_SPM.png. Accessed 18 May 2020

Morse T (2018) Twitter. Accessed 11 May 2020

Mount Pinatubo (2011) About Mount Pinatubo. http://mountpinatubo.net/. Accessed 24 April 2020

NASA (2016) Climate science investigations. In: Lambert J et al. Module "Energy, the driver of climate," Page "The greenhouse effect". http://www.ces.fau.edu/nasa/module-2/how-greenhouse-effect-works.php. Accessed 15 June 2021

NOAA (2021a) NOAA's annual greenhouse gas index. https://gml.noaa.gov/aggi/. Accessed 15 June 2021

NOAA (2021b) Coriolis. https://scijinks.gov/coriolis/. Accessed 8 March 2021

Planet for Life (2017) The history of atmospheric carbon dioxide on earth. http://www.planetforlife.com/co2history/. Accessed 5 Nov 2020

Post E et al. (2019) The polar regions in a 2 °C warmer world. https://advances.sciencemag.org/content/5/12/eaaw9883. Accessed 4 Feb 2021

Prather M et al (16 Jun 2015) Measuring and modeling the lifetime of nitrous oxide including its variability. J Geophys Res Atmos 120(11): 5693–5705. Published online 2015 Jun 5. https://doi.org/10.1002/2015JD023267

Savarino J (2015) L'ozone au cœur de la chimie atmosphérique, Futura sciences. https://www.futura-sciences.com/sciences/dossiers/chimie-ozone-coeur-chimie-atmospherique-771/page/4/. Accessed 17 April 2021

Schaefer K (2021) Methane and frozen ground. https://nsidc.org/cryosphere/frozenground/methane.html. Accessed 23 Feb 2021

Struzik E (2020) How thawing permafrost is beginning to transform the arctic. https://e360.yale.edu/features/how-melting-permafrost-is-beginning-to-transform-the-arctic. Accessed 23 Feb 2021

US EPA (2014) Climate change indicators: global greenhouse gas emissions. https://19january2017snapshot.epa.gov/climate-indicators/climate-change-indicators-global-greenhouse-gas-emissions_.html. Accessed 18 May 2020

US EPA (2016a) Climate change indicators: atmospheric concentrations of greenhouse gases. https://www.epa.gov/climate-indicators/climate-change-indicators-atmospheric-concentrations-greenhouse-gases. Accessed 18 May 2020

US EPA (2021a) Climate change indicators: climate forcing. https://www.epa.gov/climate-indicators/climate-change-indicators-climate-forcing. Accessed 18 May 2021

US EPA (2021b) Understanding global warming potentials. https://www.epa.gov/ghgemissions/understanding-global-warming-potentials. Accessed 30 May 2020

Whetstone J (2 Feb, 2021) Guest lecture presentation to John Shideler's undergraduate class on greenhouse gas management at the Rochester institute of technology. (virtual)

Chapter 2
Policy Frameworks

2.1 International Efforts to Combat Climate Change

International cooperation to combat climate change is important because countries in Europe and the USA that industrialized early are now being matched by industrialization which has occurred in China, India, and elsewhere. Disparities abound not only in terms of current contributions of greenhouse gases (GHGs)—China is now the world's largest emitter—but also in terms of cumulative contributions of GHG pollution emitted since about 1850 when industrialization accelerated, first in Europe and then in America. These realities of past contributions of GHG emissions led to the inclusion in the United Nations Framework Convention on Climate Change (UNFCCC) of the notion of "common but differentiated responsibilities" for addressing climate change.

GHG emissions are a "tragedy of the commons" type of problem. When no country or company owns a resource, such as the atmosphere, but all have access to it, management of the resource is made more difficult. Moreover, the problems associated with global warming are so daunting that only a coordinated, international response can be effective. If only a few countries try to address global GHG pollution, attempts to bend the curve on emissions will fail, and the sacrifices—economic and social—made by the few will be in vain. For this reason, efforts to combat climate change have focused on negotiating and enforcing international agreements.

Tragedy of the Commons

The "tragedy of the commons" concept is particularly applicable to climate change. The concept covers "a situation in a shared resource system where individual users, acting independently according to their own self-interest, behave contrary to the common good of all users by depleting or spoiling the shared resource." The concept first described forests or fields set aside in Europe during the Middle Ages for the common use of people living in villages.

© The Author(s), under exclusive license to Springer Nature Switzerland AG 2021
J. C. Shideler and J. Hetzel, *Introduction to Climate Change Management*,
Springer Climate, https://doi.org/10.1007/978-3-030-87918-1_2

Some farmers engaged in land-grabbing, and others failed to exercise the same kind of stewardship for lands held in common that they would for lands they owned. The concept was originally described by the British economist William Forster Lloyd during the nineteenth century and repurposed in 1968 by Garret Hardin.This concept is now applied to the care for all shared and unregulated resources such as the atmosphere, oceans, fisheries, and even the refrigerator in a shared office break room.

The principal policy mechanism for combatting climate change in the 2020s, 2030s, and 2040s is the Paris Agreement. The Paris Agreement was adopted by 196 countries in 2015 at a meeting of the UNFCCC to replace the Kyoto Protocol that was adopted in 1997. Before discussing the Kyoto Protocol and the Paris Agreement that succeeded it, it is useful to briefly review some events that shaped the modern environmental movement.

In 1976, a process accident at an Italian chemical company in Meda, Lombardy, caused the release of a large toxic cloud that triggered the evacuation of thousands of residents and was responsible for nearly 200 cases of chloracne in the human population. More than 80,000 animals died, and nearly 2000 hectares of farmland were contaminated. The nearby town of Seveso bore the main brunt of the release (Fabiano et al 2017). Subsequently, the European Union enacted the Seveso Directive that imposed long overdue industrial safety regulations on industry (Jain et al 2017).

A toxic chemical release with much more severe consequences occurred in 1984 at a Bhopal, India, chemical plant operated by a subsidiary of the American company Union Carbide Corporation. Methyl isocyanate gases released by this facility killed several thousand nearby residents and caused thousands more to flee for safety. Investigations concluded that the leak occurred due to substandard operating and safety procedures. The accident and its controversies and recriminations that followed served as a wake-up call to the chemical industry (Broughton 2005).

These two highly publicized incidents spurred legislative bodies in the 1970s and 1980s to enact new laws to control pollution and to hold companies accountable for their actions.[1] Then, in 1987 on behalf of the United Nations, the Norwegian Prime Minister Gro Harlem Brundtland issued an influential report "Our Common Future" that promoted "sustainable development." The report, which is often referred to as the "Brundtland Report," called for economic development that could meet the needs of today without compromising the needs of future generations (Fenech et al 2003).

As a result of numerous prominent environmental issues arising in the 1970s and 1980s, and a growing scientific concern about the risks associated with climate change, policy makers convened a global summit to agree on a coordinated response. In 1992, world leaders gathered in Rio de Janeiro, Brazil, for the United Nations

---

[1] For example, fires occurring on the Cuyahoga River in Ohio in the 1970s prompted a major revision to the U.S. Clean Water Act. The Resources Conservation and Recovery Act passed the U.S. Congress after news emerged about the impacts of toxic waste dumped at the Love Canal (NY) landfill.

Conference on Environment and Development, often later referred to as the "Rio Summit." There they signed the Rio Declaration on Environment and Development and created the United Nations Framework Convention on Climate Change. The convention, which had the status of an international treaty, became effective in March 1994.

2.1.1 The UNFCCC

Under the convention countries agreed to meet annually in a "Conference of the Parties" to further develop international cooperation on climate change. These meetings are often referred to informally as "COPs." The 1997 COP in Japan produced the Kyoto Protocol which was the first international effort to reduce GHG emissions by common action. From 2021, the Kyoto Protocol was replaced by the Paris Agreement negotiated in 2015 at COP21 in Paris. This agreement required participating countries to submit to the UNFCCC their "nationally determined contributions" (NDCs) to reduction of GHG emissions. Initial NDCs were due by 2020 and revisions to them every five years thereafter (i.e., by 2025, 2030, etc.). From its beginning, a Subsidiary Body for Scientific and Technological Advice operated within the UNFCCC to provide participating countries with timely information and advice on scientific and technological matters as they related first to the Kyoto Protocol and later to the Paris Agreement.

2.1.2 The Kyoto Protocol

The Kyoto Protocol was the first international agreement targeting the reduction of greenhouse gas emissions. Although USA's Vice President Al Gore had played a prominent role in its negotiation, President Bill Clinton did not submit the Kyoto Protocol to the U.S. Senate for approval due to overwhelming opposition in that body to its adoption. It went into effect anyway after being ratified by 147 countries. The Russian Federation's ratification in 2004 ensured its entry into force (United Nations 2004).

The Kyoto Protocol enshrined the principle of "common but differentiated responsibilities."[2] This meant that the countries whose industrialization and abundant use of fossil fuels had most contributed to the buildup of CO_2 concentrations in the atmosphere should shoulder the most responsibility for their abatement. Key to implementation of the Kyoto Protocol was three flexible mechanisms: emissions

---

[2] This principle recognized that while developed countries bore disproportionate responsibility for the emissions that caused climate change, all countries shared a common responsibility to work to mitigate greenhouse gas emissions. It was first articulated in the United National Framework Convention on Climate Change in 1992.

trading, a Clean Development Mechanism (CDM), and Joint Implementation (JI). The flexible mechanisms were inspired by the Clean Air Act Amendments in the USA which had already shown remarkable success in reducing acid rain emissions from US power generation plants. CDM was managed by an Executive Board which adopted emission reduction and removal enhancement methodologies and approved the third-party bodies that validated and verified them.

Under the Kyoto Protocol, advanced industrialized countries, identified in Annex I of the protocol, were assigned targets for reducing their emissions compared to 1990 levels. The targets were a result of political negotiation. States in the former Union of Soviet Socialist Republics (USSR), whose economies had collapsed after the union of republics disbanded in 1991, received allocations of emissions far in excess of their post-USSR needs. Australia, which was a growing exporter of coal to trading partners in Southeast Asia, was allowed to grow its emissions by 8% during the first commitment period (2008–2012). In 1997, China was considered a developing country and was not included in the Annex I list. The burden of real emission reductions fell most on countries in the European Union, Japan, and Canada.

Annex I parties were industrialized members of the Organisation for Economic Co-operation and Development (OECD) in 1992, plus countries with economies in transition including the Russian Federation, the Baltic states, and several Central and Eastern European states.

2.1.3 The Clean Development Mechanism (CDM)

The CDM allowed developed, industrialized countries to take credit for reducing emissions in developing countries whose emissions were not capped. JI allowed companies to finance the reduction of emissions in other developed countries, such as Russia and Ukraine. For their investments, Annex I countries, or companies within them, received CDM and JI credits to use in meeting their own emissions targets. In the first two phases of the European Union Emission Trading System (EU ETS), regulated installations surrendered almost equal numbers of CDM's Certified Emission Reductions (CERs) and JI's Emission Reduction Units (ERUs) to meet their compliance obligations. Total offsets amounted to just over 1 billion CERs and ERUs[3] (Głowaki 2013).

CDM projects appear to be numerous—12,516 submitted projects—but only 63% (7,854) were registered.[4] By 2020, 4,720 registered projects were considered dormant

---

[3] In 2014 for the 2013 reporting year, the largest number of surrendered credits originated from projects in China, Ukraine, and Russia. Eighty percent of CERs originated from China, and 95 percent of ERUs originated from projects in Ukraine (70%) and Russia (25%).

[4] "Registration" of a project occurs when the Executive Board of the CDM accepts a validation report from a "designated operational entity" (DOE). This process step makes the project eligible to submit verification reports demonstrating the achievement of emission reductions. CERs are issued by the CDM Executive Board after it reviews and accepts DOE-submitted verification reports.

since they had not had contact with the UNFCCC Secretariat since 2014. At the end of 2020, only 3,125 registered projects remained active (CDM 2021).

CDM projects were developed in a limited number of countries. These included:

- China, 4977 projects
- India, 3139 projects
- Brazil, 740 projects
- South Africa, 103 projects
- Israel, 57 projects
- Saudi Arabia, 7 projects
- Fiji, 7 projects (not registered)
- Haiti, 5 projects (not registered).

The number of credits granted for each project varied considerably. The largest five projects produced more than one million metric tons per year, while the average project contributed about 130,000 tons of emission reductions (CDM 2021).

The largest source for surrendered CDM offset credits in the EU ETS after the linking directive came into effect in 2008 was refrigerant manufacturing companies in China. These were allowed, under the Montreal Protocol, to continue manufacture of HCFC-22 long after bans on the manufacture of this chemical resulted in its phaseout in developed economies. A co-product of the manufacture of HCFC-22 was HFC-23, a high-GWP chemical with no commercial use (Miller and Batchelor 2012). CDM projects financed the destruction of this gas to prevent its emission to the atmosphere. The different phaseout schedules for HCFC-22, then, were the root cause for the continued manufacture of this refrigerant and its resulting co-product.

Even worse, the economics of destroying HFC-23 were such that many observers believed that manufacturers were incentivized to produce more HCFC-22 so they could receive CDM offset credit money for destroying its useless co-product. The circumstances surrounding the issuance of these carbon credits caused the European Commission to disqualify CERs from this source during the third phase of the EU ETS. The European Commission also restricted eligible projects to those in the "least developed countries" (LDCs) and limited carbon credit use to 4.5% of regulated installations' verified emissions (Głowacki 2013).

Joint Implementation covered projects located in countries of the former Soviet Union, such as Russia and Ukraine. JI contributed far fewer projects than the CDM, but this Kyoto mechanism was marked by a lack of transparency and produced a glut of ERUs of "highly questionable environmental integrity" (Carbon Market Watch 2013).

There are two types of JI projects:

- Track 1 projects are approved, and credits are issued by the host countries themselves.
- Track 2 projects are approved by the JI Survey Committee (JISC), an international entity similar in structure and function to the CDM Executive Board.

For every JI offset issued, the host country must cancel one of its Kyoto allowances (AAUs) to avoid the double counting of emissions reductions. This has been used as

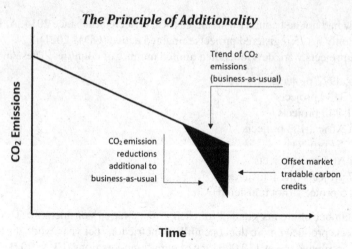

Fig. 2.1 Principle of additionality (graph redrawn and translated from French original in Criqui et al. 2009)

a reason to argue that for JI projects, concerns about additionality—the requirement that projects represent genuine emissions reductions that would not have happened anyway—are not relevant. This however is not true. More than 90% of the credits have been issued by Russia and Ukraine, both countries with a very large AAU surplus (so-called "hot air".[5] Effectively, these countries have an incentive to inflate the stated emissions reductions because they have so many spare AAUs that they could meet their own emission targets anyway, even if they issued a very large number of JI credits (Fig. 2.1).

2.2   The Paris Agreement

The Paris Agreement was negotiated at the twenty-first COP to provide a policy framework to replace the Kyoto Protocol. Signatory countries agreed to target emission reductions that would limit global warming to no more than 1.5 to 2 °C above pre-industrial levels. Negotiators knew that required emission reductions could not be realized if the world's top two greenhouse gas polluters—China and the USA—remained outside the agreement. US negotiators recognized the difficulty of obtaining ratification in the U.S. Senate if the new agreement was framed as an international treaty, so they insisted that compliance to the protocol should remain voluntary.

---

[5] The base year for allocation of Assigned Amount Units was determined by Kyoto negotiators to be 1990—the year prior to the economic collapse of the former Soviet Union. As a result, the newly independent republics that succeeded the USSR were given AAUs far exceeding their economic needs. Environmental NGOs quickly tagged this overhang of AAUs as "hot air" (Carbon Market Watch 2013).

The foreword to the Paris Agreement noted the scientific consensus that the global average increase in temperature of 1 °C since 1880 resulted from an increase in emissions of GHGs and that the greenhouse effect was mainly attributable to human activities. The Paris Agreement enshrined the following objectives:

1. To limit global warming to an increase between 1.5 and 2 °C by 2050
2. To establish multiparty financing processes (states, regional financing institutions, banks, and the private sector) from 2020, funded at $100 billion per year, to help developing states mitigate their GHG emissions, notably through activities financed by the Green Climate Fund (2021) that was established per the Lima Agreement (reached at COP20) in Seoul, South Korea.

The Paris Agreement called upon the world's countries to voluntarily reduce greenhouse gas emissions through NDCs that are reviewed every five years (UNFCCC 2015). At each review period, the ambition of these NDCs is to be ratcheted up as needed to meet the Paris Agreement objectives.

2.3 The European Union's Emissions Trading System

To honor their commitments under the Kyoto Protocol, European states implemented the EU Emissions Trading System, a cap-and-trade program. The EU allocated emission allowances, called Assigned Amount Units, to countries in accordance with their percentage shares of the 1990 baseline negotiated in the Kyoto Protocol. Each country in the system then provided European Union Allowances (EUAs) to companies operating in their jurisdictions. In the first and second phases (2005–2007 and 2008–2012), countries provided EUAs to facilities at no cost. This policy was intended to protect European industry from unfair competition from industries in countries that did not face a carbon reduction mandate. Auctioning of allowances became the default distribution method in the third phase (2013–2020), with harmonized rules for those allowances that were still given away for free. The "cap" on emissions incentivized companies to become more energy efficient, so they could increase production and sales without paying more for the right to emit GHGs. Between 2005 and 2019, GHG emissions from the facilities covered by the ETS declined by approximately 35% (European Council 2020).

EU ETS

The European Union Emissions Trading System (EU ETS), through the member states, imposes a cap on emissions from industrial facilities. It covers approximately 10,000 installations in the sectors of electricity generation, central heating networks, steel, cement, oil refining, glass, paper, etc., which account for more than 40% of European greenhouse gas emissions. Allowances equal to the cap are distributed each year. After the end of each year, facilities are required to surrender a quantity of allowances equivalent to their actual emissions.

A facility that emits more than its allocation must purchase additional allowances: This is the "polluter pays" principle. A facility that emits less than its allocation can sell unused allowances and bank the revenues. These can be mobilized, for example, to finance investments allowing them to control their emissions. The ETS Directive provides that member states should use at least 50% of auctioning revenues or the equivalent in financial value for climate- and energy-related purposes.

The proceeds from the EU's allowance auctions are allocated to each state which report annually the distribution financing for energy renovation of buildings. Around 80% of revenues received by member states in 2013–2018 were used for climate- and energy-related purposes. Member states report annually on the amounts and use of the revenues generated.

Under the Monitoring Mechanism Regulation, covered companies may buy and sell allowances on the European emissions trading market. Covered facilities that do not meet the cap with free and purchased EUAs can purchase CERs on the market to meet their compliance obligations.

The first phase (2005–2007) of the EU ETS allowed European states and companies to prepare for the first commitment period under the Kyoto Protocol. Two phases (2008–2012 and 2013–2020) were managed according to the rules of the Kyoto Protocol. The fourth phase (2021–2030) was aligned to the requirements of the Paris Agreement.

A 2009 directive extended the scope of the scheme and changed the allowance allocation system for the third phase (2013–2020). The system was modified again after the Paris Agreement in line with Europe's ambition to reduce its GHG emissions 40% compared to the 1990 baseline. This revision for the fourth phase focused on:

– Strengthening the EU ETS as an investment driver by increasing the pace of annual cap reduction to 2.2% as of 2021 and reinforcing the Market Stability Reserve (the mechanism established by the EU to reduce the surplus of emission allowances in the carbon market and to improve the EU ETS's resilience to future shocks)
– Continuing the free allocation of allowances as a safeguard for the international competitiveness of industrial sectors at risk of carbon leakage, while ensuring that the rules for determining free allocation are focused and reflect technological progress
– Helping industry and the power sector meet the innovation and investment challenges of the low-carbon transition via dedicated funding mechanisms—the Innovation Fund and Modernization Fund.

(European Commission, 2021a).

At the time the EU ETS was implemented, participants in the cap-and-trade program assumed that countries would have a difficult time meeting their Kyoto targets. In early 2006, the value of EUAs rose to €30/t from €7/t in 2005 as companies traded cash for allowances ahead of published reports of GHG inventories. However, on 28 April 2006 France and Spain announced unexpectedly low annual emissions of CO_2 during the first reporting year. The price for EUAs crashed as a result, from €30.60 on 24 April 2006 to €11.50/t on 2 May 2006. Following the release of more emissions data on 12 May 2006, EUA prices dropped even lower to €8.60. Markets had concluded that allowances in the first phase of the EU ETS were overallocated, particularly in the electricity generating sector. But uncertainty about that question caused a rebound in EUAs to nearly €20t by the end of May 2006. The market collapse in EUAs also affected the price of Certified Emission Reductions (CERs) from CDM projects. Their value dropped even more, trading in 2006 as low as €6.80/t. They would fall even further in later phases of the EU ETS market.

Overallocation of free EUAs occurred in Europe for two reasons. The first reason is that policy makers, when negotiating where to set the cap on emissions, sought to protect existing companies in their countries from large adverse economic impacts. They kept in mind the importance of the companies to the local economy, to tax revenues, and to employment, and worked to inflict the least possible economic pain on them while still adhering to their legal obligations under the Kyoto Protocol. The second reason the caps were set too high is that policy makers underestimated the ability of companies to reduce their emissions of CO_2. Once the EU ETS put a price on these emissions, companies looked for ways to reduce them. The degree of their success in doing so was unanticipated, just as the ability of US companies to reduce emissions of sulfur dioxide and oxides of nitrogen under the rules of the Clean Air Act Amendments of 1995 had been underestimated.

Among the covered industries, annually reported GHGs did not significantly exceed allocated amounts in the first phase of the EU ETS (2005–2008) and in the Kyoto Protocol's first commitment period (2008–2012). The Great Recession (2008–2009) is a confounding factor; some analysts believe that reductions in GHG emissions to the end of the Kyoto first commitment period were due more to the effects of the recession than to the influence of the ETS (Millington et al 2020 pp 31–32). In the second Kyoto commitment period, 2013–2020, the EU lowered the cap on emissions. As a result, the price of EUAs gradually rose and traded in a range of €20–30/t in 2020, the last year of the second Kyoto commitment period. CERs, however, traded at less than €0.50/t as 2020 came to a close. The difference in price reflected the large oversupply of CERs available on the market and the limitations on their convertibility into EUAs imposed by the EU in its third phase.

The price of emission reductions matters, since most experts consider that only a minimum price of €100 per ton of CO_2 would incentivize rapid decarbonization of the economy. The price of AAUs increased in 2018–2019 to a range between €10 and €30 per ton, recovering price levels it had attained during the first phase of the EU ETS. In 2021, AAU prices soared to more than €60 per ton.

In the first decade of the twenty-first century, the EU set ambitious goals for emission reductions, including a "firm and unilateral commitment to reduce by at least

20% by 2020." It qualified this goal however by linking it to the conclusion of a global multilateral agreement for the third stage (2013–2020) of the EU ETS (EU Council 2007). It further committed to a 30% reduction by 2020 provided that other developed countries committed to comparable targets and advanced developing countries made contributions tailored to their national circumstances. Policy makers clearly had in mind the hoped-for success of COP15 in Copenhagen in 2009 and expected favorable outcome of renegotiations of the second Kyoto commitment period. They would be disappointed at Copenhagen, but continued negotiations finally led to the Paris Agreement in 2015.

Despite difficulties encountered, the EU ETS delivered on its goal of reducing emissions 20% by 2020. According to the European Commission, EU greenhouse gas emissions were reduced 24% between 1990 and 2019, while the economy grew by around 60% over the same period (European Commission 2021b).

2.4   Cap-and-Trade in the USA

Because the USA did not ratify the Kyoto Protocol and never participated in its flexible mechanisms for regulated industries, it was left to private initiative and the states to establish cap-and-trade programs. First to market was the Chicago Climate Exchange, followed by the Regional Greenhouse Gas Initiative, and then the California Air Resources Board, an agency of California government.

The Chicago Climate Exchange was a voluntary, private-sector initiative that featured two phases, 2003–2006 and 2007–2010. Members committed to reduce their emissions of the six Kyoto gases by 4% in Phase I and by 6% in Phase II. Each member received free allocations of allowances equal to their baseline emissions which were derived from a four-year average of their emissions from 1998 through 2001 or those reported for the year 2000. In addition to achieving emission reductions in their operations, members could apply carbon credits from projects located in the USA , Canada, Mexico, and Brazil to meet emission reduction targets. The program attracted about 40 companies who made legally binding commitments to adhere to the program rules (Kerr 2007).

2.4.1   The Regional Greenhouse Gas Initiative

The Regional Greenhouse Gas Initiative, known as RGGI (pronounced "reggie"), was created by agreement in 2005 among the governors of seven northeastern and mid-Atlantic states. It was the first regulatory cap-and-trade program in the USA, covering only electricity generating facilities. A "model rule" adopted in each state in 2006 established a regional CO_2 emissions budget. The cap-and-trade program went into effect in 2009 to regulate CO_2 emissions from fossil fuel-fired electricity generating units having a rated capacity equal to or greater than 25 megawatts. Three additional states subsequently joined the initiative, and in 2020, Virginia became its latest member (RGGI 2021a).

The initial carbon budget for electricity generation, which had been based on estimated future emissions, exceeded actual emissions when the cap first came into effect. This resulted in an overallocation of RGGI allowances to electricity generating facilities. During the first years of the program, RGGI compliance allowance prices traded in a fairly low range and only began to rise significantly after the adoption of amendments to the program in 2017. The model rule provided for the use of offset credits, but no projects were developed under RGGI rules until 2017 when a methane reduction project in Maryland became the first project to produce RGGI offsets (3Degrees 2017). Allowance prices languished in the $2.00 or less range from 2010 through 2012, and then started rising. Allowance prices in late 2020 cleared above $7.00 for only the second time since auctions began in 2008 (RGGI 2021b) and surged in 2021 to more than $11.00.

2.4.2 California's Climate Action

If it were a country, California would be the world's fifth largest nation by size of its economy. The state's economy features a large aerospace industry, prime agricultural lands, and with Los Angeles, the second largest urban area in the United States after New York. With its industrialization, growing population, and dependence on automobiles for transportation, California by the mid-twentieth century faced serious air quality issues, particularly in southern California. In response, the state in 1967 established the Air Resources Board, an agency devoted to clearing the air of smog-producing pollution. Several decades later, the California legislature extended the agency's mission to rolling back GHG emissions to 1990 levels, a reduction of about 14% from the 490 million tons recorded in 2004[6] (CARB 2021).

The "Global Warming Solutions Act of 2006"—more commonly referred to as "AB 32"—called for regulations that would ensure the reduction of greenhouse gas emissions through energy efficiency, the use of renewable energy, clean transport, and waste reduction. Its various sections provided authority for actions in the following areas:

1. A cap-and-trade carbon trading market, with a declining cap, trading of allowances, and a role for carbon credits from approved project types
2. Reducing emissions from public lighting
3. The development of electric vehicles with low greenhouse gas emissions
4. Energy efficiency
5. The development of the use of renewable resources
6. The development of low-carbon energy

---

[6] According to Assembly Bill 32, the "Global Warming Solutions Act of 2006," CARB was directed to prepare scoping plans to achieve the maximum technologically feasible and cost-effective GHG emission reductions (California Air Resources Board by 2020).

7. Regional transport with greenhouse gas reduction targets
8. Energy-efficient vehicles
9. The transport of goods
10. A million square-meter solar roof program
11. Large and heavy carrier vehicles
12. Industrial emissions
13. High-speed trains
14. The green building strategy
15. Gases with high greenhouse gas potential
16. Waste recycling
17. Sustainable forests
18. Water
19. Agriculture.

The results of the climate mitigation initiatives launched by AB 32 have been remarkable. During the period 2000 to 2018, California's per capita GHG emissions declined from a peak in 2001 from 14 tons per person to 10.7 tons per person in 2018, a drop of 23%. In the same period, the carbon intensity of California's economy—the amount of carbon pollution per million dollars of gross domestic product (GDP)—declined 43% from the 2001 peak, while the state's GDP increased by 59% over this period. The state's goal of reducing its GHG emissions to 1990 levels by 2020—enshrined in AB 32—was reached in 2017. In 2017, California set a 2030 climate target to reduce GHG emissions by 40% from 1990 levels and to advance substantially toward a goal to reduce GHG emissions to 80% below 1990 levels by 2050 (CARB 2017). Figure 2.2 shows that over this period of increasing population and economic activity absolute GHG emissions declined by 12% over peak 2001 emissions.

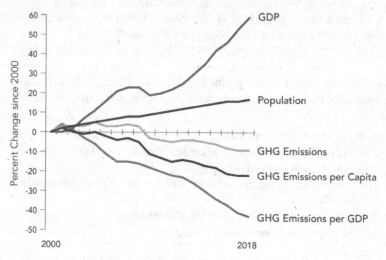

Fig. 2.2 Change in California GDP, population, and GHG emissions 2000–2018 (CARB 2021)

2.4.3 California Cap-and-Trade Program

California's cap-and-trade program, which launched in 2013, was the first multisectoral cap-and-trade program in North America. Approximately 450 businesses in California responsible for about 85% of California's industrial GHG emissions are enrolled in the program. The threshold for participating in the program is 25,000 t/CO_2e emissions per year for most sectors. Regulated industry received allowances through a combination of free distribution and quarterly auctions, with a majority sold at auction (C2ES 2021). A reserve price of $10/t was set by regulation for 2013 allowances sold in 2012, and it increased by 5% per year plus the rate of inflation. The California program was linked with Quebec's cap-and-trade program in 2014, and since then, the two jurisdictions have conducted joint auctions.

California learned several lessons from the implementation of cap-and-trade in Europe and from RGGI in the eastern USA. Like the EU ETS, but unlike RGGI, California's program is broadly based with respect to economic sectors. Unlike RGGI, California based its cap for the first compliance period on actual emissions reported by industry and verified by third parties starting in 2009. With measured emissions data, California avoided overestimating industry's GHG emissions as the administrators of RGGI had done. California also set a meaningful price floor for the auction of allowances, another failure of program design by RGGI. California did not allocate as many free allowances as the EU ETS and RGGI programs had. Finally, California's registry avoided problems with fraud that plagued the EU ETS which arose due to weak administrative and decentralized controls (McDowell 2017).

2.5 California Low-Carbon Fuel Standard

California's AB 32 also targeted greenhouse gas emission reductions from the production, transportation, and use of road transportation fuels. The first goal was to reduce these 10% by 2020 from a 2010 baseline. To achieve this, in 2009 the Air Resources Board created a new regulatory program called the "low-carbon fuel standard" (LCFS). The LCFS was designed to decrease the carbon intensity of California's transportation fuels and to increase the availability of low-carbon and renewable alternatives. As Californians in 2013 used 14.2bn gallons of gasoline and 2.7bn gallons of diesel fuel, the program had the potential to make a significant impact on the state's CO_2 emissions.

Carbon intensity, or "CI," is calculated for each fuel used for transportation on a life cycle basis considering emissions from all phases of the fuel's production up to its combustion in vehicles. The regulation requires suppliers of transportation fuels to demonstrate that the mix of fuels they sell in the California market meets the LCFS intensity standards during each annual compliance period. These standards required CI reductions of 2% in 2016, with rates of reduction increasing each year until 2020 when a 10% reduction would be achieved (CARB 2015).

Under the LCFS, fossil fuel refineries generate "deficits" for each gallon of fuel sold. On the other hand, ethanol producers, whose fuel was blended with gasoline, earn "credits" for their product which is made from renewable sources. Suppliers of transportation fuel must ensure that the number of credits they earn (or acquire in the marketplace from other parties) is equal to, or greater than, the deficits they incur. Deficits and credits are expressed in tons of GHG emissions. Suppliers report fuel transactions annually to ARB along with their determination of the number of deficits and credits generated. Credits may be banked or traded within the LCFS market to meet obligations (CARB 2015).

The LCFS defines "carbon intensity" standards for gasoline and diesel fuel. Their renewable substitutes, like biofuels, and alternates, like electricity and hydrogen, usually obtain calculated CIs based on specific life cycle emission "pathways" that count GHG emissions from farm or other source of feedstock to the final fuel processing plant. Life cycle assessment techniques quantify the amounts of carbon dioxide (CO_2), methane (CH_4), nitrous oxide (N_2O), and other greenhouse gases emitted during the fuel production cycle from well or field to the wheels of the vehicles. For some biofuels, the CI includes a calculation of indirect emissions from land use change. The overall GHG contribution from all production and transportation steps is divided by the fuel's energy content measured in megajoules. Thus, carbon intensity is expressed in terms of grams of CO_2 equivalent per megajoule (gCO_2e/MJ). The LCFS is performance-based and fuel neutral, allowing the market to determine how the carbon intensity of California's transportation fuels will be reduced (CARB 2015).

According to ARB LCFS staff analysis, reporting parties through mid-2014 had cumulatively generated about 8.7 million metric tons of LCFS credits and 5.2 million metric tons of deficits, for a net total of about 3.5 million metric tons of "excess" credits. This early overcompliance by regulated parties built up significant surplus credits that could be used for future compliance as fuel carbon intensity requirements tightened. The primary source for credits under the LCFS came from ethanol (60%), with other fuels contributing lesser amounts: renewable diesel (15%), biodiesel (13%, natural gas (10%), and electricity (2%) (CARB 2015).

California regulators calculated the carbon intensity of certain standard transportation fuels and assigned them "default" carbon intensity values. But it also permitted fuel producers to submit customized life cycle assessments in case a producer could demonstrate a lower life cycle carbon intensity number. This alternative submission method during the first five years of the program resulted in more than 230 new or modified fuel "pathways" with substantially lower carbon intensity numbers accepted by regulatory authorities. Almost 170 biofuel facilities are registered under the LCFS as supplying low-carbon fuels to California (CARB 2015).

ARB estimated that the LCFS will reduce transportation-related GHG emissions by 35 million tons during the period 2016–2020. The program functions much like an emissions reduction trading program in which credits are awarded based on the performance of fuels that exceed a regulatory standard. Credits can be banked indefinitely until they are sold and transferred, exported to other programs, or retired for compliance purposes (CARB 2015). In 2021, LCFS credits traded in a range from

$160/t to $200/t, a level that provided a strong price signal to anyone who could deliver a low-CI transportation fuel to the California market.

California's LCFS is widely viewed as a regulatory success. Similar regulations have been adopted in Oregon (Oregon Secretary of State 2019) and British Columbia (2020), and several other US states are considering establishing their own programs. At the federal level, Environment and Climate Change Canada has published regulations to implement a Clean Fuel Standard (CFS) that would take effect in 2022 (ECCC 2021).

The state of California has set many air quality, climate, and community risk reduction goals for the thirty-year period 2020–2050. The state's Air Resources Board (CARB) is developing the 2020 Mobile Source Strategy to achieve California's targets. Mobile sources—and the fossil fuels that power them—continue to contribute a majority of diesel particulate matter emissions as well as smog- and particulate-forming pollutants such as oxides of nitrogen (NOx) and the largest portion of greenhouse gas emissions in California. California Senate Bill 44, signed by Governor Newsom on 20 September 2019, acknowledged the substantial contribution that mobile sources make to statewide emissions and required CARB to update the Board's 2016 Mobile Source Strategy by 1 January 2021 and every five years thereafter. The actions contained in the Mobile Source Strategy aim to deliver broad environmental and public health benefits, as well as support much needed efforts to modernize and upgrade transportation infrastructure, enhance system-wide efficiency and mobility options, and promote clean economic growth in the mobile sector (CARB 2020).

2.6 Regulation of GHG Emissions in Canada

Canada, whose greenhouse gas emissions represented just 2% of the world's total in 1990, signed and ratified the 1997 Kyoto Protocol. In an era of expanding shale gas research and development and exploitation of high GHG intensity oil sands in Alberta, the federal government was not able to establish a mechanism, such as a nationwide cap-and-trade program, to achieve emission reductions in line with its Kyoto commitment. In December 2011, the federal government announced its withdrawal from the agreement, citing the lack of participation by the USA and China (Guardian 2011). By this time, early actions to regulate Canadian GHG emissions had been enacted at the provincial level in Quebec, Ontario, Alberta, and British Columbia. In 2015, Canada embraced the Paris Agreement and the federal government took steps to ensure that aggregate emission reductions undertaken at the provincial level would meet the level of ambition outlined in Canada's NDCs submitted to the UNFCCC.

2.6.1 Ontario's Greenhouse Gas Reduction Plan

Ontario is Canada's most populous province. Its economy alone comprises 38.6% of the gross domestic product of Canada (Ontario Ministry of Finance 2021). Ontario adopted a mandatory reporting of GHG emissions regulation in 2009 that required industry to begin reporting GHG emissions from 2010. A 2016 regulation created a cap-and-trade program that took effect on 1 January 2017. Midway through its first year of its implementation, however, Ontario voters selected a new premier who opposed the cap-and-trade policy and repealed it (CBC 2018). The province maintained mandatory GHG emissions reporting for companies emitting more than 25,000 tons of CO_2e. As a result of ending cap-and-trade, the federal government found that Ontario's climate efforts were insufficiently robust and designated the province as subject to the backstop Output-Based Pricing System (OBPS) for fuel purchased for industrial use. Ontario's participation in this program became effective on 1 January 2019.

Later in 2019, Ontario adopted a new emission performance standards regulation (Ontario 2019) whose intent was to substitute an Ontario-administered program for the federal OBPS. Environment and Climate Change Canada (ECCC) accepted this regulation of industrial facilities as equivalent to the OBPS in September 2020. The effective implementation date of this regulation will be determined by the date by which ECCC releases Ontario from its obligation to participate in the federal OBPS program (Ontario 2020).

2.7 Alberta's Regulation of GHG Emissions

Alberta is Canada's number one province for oil and gas development, producing 80% of the country's crude oil. Canada is the fourth largest producer and third largest exporter of oil in the world. The USA is Canada's largest customer, purchasing 98% of Canada's oil exports. Nearly two-thirds of Canada's oil production is sourced from Alberta's oil sands (Natural Resources Canada 2020).

The province of Alberta introduced a Climate Change Strategy in 2007 that targeted GHG emissions intensity reductions through the implementation of two regulations. From 2005, the Specified Gas Reporting Regulation required all Alberta facilities whose annual GHG emissions exceeded 50,000 tons to report their emissions to the Alberta Environment and Sustainable Resources Development (AESRD). In accordance with Alberta's Specified Gas Emitters Regulation (SGER), facilities emitting more than 100,000 tons CO_2e annually were regulated by AESRD and were required to reduce GHG emissions intensity by 12% per production unit from an approved baseline. Compliance with the program could be achieved either by facilities making absolute emission reductions to align their emissions intensity with the target, by purchasing Alberta offset credits or Technology Fund credits or by using banked Emission Performance Credits from prior years (IETA 2015).

On 1 January 2018, the Alberta government introduced its Carbon Competitiveness Incentive Regulation (CCIR) to replace the SGER. The CCIR uses product-based benchmarks to establish emissions limits above which compensation is required. Compensation mechanisms may include generation of emissions performance credits and use of credits or offsets, as well as payment to Alberta's Climate Change and Emissions Management Fund. Product emission benchmarks are to be reviewed every five years. This new regulatory approach was accepted by ECCC which exempted Alberta from the backstop provisions of the OBPS (Alberta 2018).

2.7.1 British Columbia's Climate Change Strategy

British Columbia (BC) pioneered a carbon tax on motor and heating fuels that took effect in July 2008 at a rate of C$10 tCO$_2$e. The tax was initially "revenue neutral," meaning that other taxes were reduced to compensate taxpayers for the rises in fuel prices (McInnnes 2008). By 2019, the tax had increased to C$40 tCO$_2$e and the government abandoned its commitment to revenue neutrality while still providing income tax rebates to BC residents.

Through its Greenhouse Gas Reduction (Cap-and-Trade) Act, which passed in 2008, BC required industrial facilities emitting 10,000 tons or more of GHGs to submit annual reports beginning with the 2010 emissions year. Cap-and-trade was not implemented, and the act was replaced in 2015 by the Greenhouse Gas Industrial Reporting and Control Act which established performance standards for industry sector emissions. BC required companies who did not meet emission performance standards to purchase offsets as compensation (British Columbia 2021a).

Through the Greenhouse Gas Reductions Target Act, enacted in 2007, British Columbia required 128 public sector organizations (municipalities and district governments) to become "carbon neutral." District governments were the primary operators of landfills, and most of these were required to collect and combust landfill gas starting in 2016, leading to some permanent emission reductions. By 2020, the province noted that public sector organization (PSO) emissions were approximately 6.5% lower than in 2010, the first year of "carbon neutrality." In 2019, in addition to their absolute emission reductions below the 2010 baseline, PSOs in 2019 reported that 651,543 tons of emissions from a total of 762,168 tCO$_2$e public sector emissions had been offset.

A "Climate Action for the 21st Century" report, published in 2010, outlined the province's goals: achieve a 33% reduction from 2007 levels of greenhouse gas emissions by 2020, with an 80% reduction goal for 2050, application of the first broad-based revenue-neutral carbon tax, 93% clean electricity production, and 100% carbon neutral government by 2010.

While British Columbia did enact North America's first carbon tax, it struggled over the ten years following 2008 to meet its greenhouse gas emission reduction targets. An updated report in 2020 acknowledged that economic and population growth had contributed by 2018 to a rise in greenhouse gas emissions of 3.5 million tCO$_2$e over 2007 levels. On the other hand, the carbon intensity of BC's economy had decreased 16% since 2007 and GHG emissions per capita had declined 8% since 2007 (CleanBC 2020).

The results achieved in British Columbia after more than a decade of effort are sobering. Early progress made in reducing the carbon intensity of the province plateaued around 2015, while at the same time the declines in per capita emissions reversed direction in 2015 and began to climb. Given that the world's carbon budget for achieving a rise of no more than 2 °C by 2050 is based on absolute emission reductions rather than relative ones, it is apparent that even more ambitious climate action will be required to achieve BC's goal of reducing GHG emissions 40% by 2030 compared to 2007 levels.

2.7.2   China

As described above, the first markets, carbon markets, were held in Chicago, Europe, the northeastern USA, California, and Quebec. A carbon emission rights market was established on the Shanghai Stock Exchange as a pilot project, in 2017. China's national carbon market started in February 2021 and operates in a similar fashion to other national cap-and-trade markets. Industries are allocated allowances for their emissions, and governing law establishes rules for trading, emissions monitoring, and verification (Slater 2021).

2.8   United Nations Sustainable Development Goals

In 2015—the same year that saw the adoption of the Paris Agreement—the United Nations adopted a set of Sustainable Development Goals (SDGs) as part of its "2030 agenda for sustainable development" (United Nations 2021). The 17 SDGs built upon many previous UN initiatives including the eight Millennium Development Goals to reduce extreme poverty by 2015. In the years since their adoption, the SDGs have been widely cited in the context of new development, standardization, assistance to developing countries, and other initiatives. The 17 goals are:

Goal 1. End poverty in all its forms everywhere
Goal 2. End hunger, achieve food security and improved nutrition and promote sustainable agriculture
Goal 3. Ensure healthy lives and promote wellbeing for all at all ages
Goal 4. Ensure inclusive and equitable quality education and promote lifelong learning opportunities for all
Goal 5. Achieve gender equality and empower all women and girls
Goal 6. Ensure availability and sustainable management of water and sanitation for all

(continued)

(continued)

Goal 7. Ensure access to affordable, reliable, sustainable and modern energy for all
Goal 8. Promote sustained, inclusive and sustainable economic growth, full and productive employment and decent work for all
Goal 9. Build resilient infrastructure, promote inclusive and sustainable industrialization and foster innovation
Goal 10. Reduce inequality within and among countries
Goal 11. Make cities and human settlements inclusive, safe, resilient and sustainable
Goal 12. Ensure sustainable consumption and production patterns
Goal 13. Take urgent action to combat climate change and its impacts
Goal 14. Conserve and sustainably use the oceans, seas and marine resources for sustainable development
Goal 15. Protect, restore and promote sustainable use of terrestrial ecosystems, sustainably manage forests, combat desertification, and halt and reverse land degradation and halt biodiversity loss
Goal 16. Promote peaceful and inclusive societies for sustainable development, provide access to justice for all and build effective, accountable and inclusive institutions at all levels
Goal 17. Strengthen the means of implementation and revitalize the Global Partnership for Sustainable Development

2.9   International Aviation and Shipping

International aviation and shipping were not included in the Kyoto Protocol because their activities were "extraterritorial," subject to international regulation rather than national regulation. This exemption from regulation stoked controversy when Europe included aviation within the EU ETS. European airplane operators complained that foreign competitors located outside the EU had an unfair advantage because they were not subject to the EU ETS. In 2012, the EU attempted to extend the EU ETS to non-EU air carriers who landed and took off from European airports, but foreign operators successfully resisted regulation by the EU. A compromise settlement deferred EU regulation as long as the body that regulated international civil aviation agreed to implement worldwide regulation of aircraft CO_2 emissions from international flights.

2.9.1   The Carbon Offsetting and Reduction Scheme for International Aviation (CORSIA)

In response to pressure from the EU, the International Civil Aviation Organization (ICAO) in 2016 approved the Carbon Offsetting and Reduction Scheme for International Aviation (CORSIA). CORSIA established mandatory annual reporting of airplane emissions from 2019 and an offsetting program from 2021 that allows airplane operators to use emissions units from carbon markets to offset the CO_2 emissions that cannot be reduced through technological and operational improvements or sustainable aviation fuels. Rules for CORSIA are defined in Standards and Recommended Practices issued by the International Civil Aviation Organization as Annex 16 to the Convention on International Civil Aviation, Environmental Protection, Volume IV, Carbon Offsetting and Reduction Scheme for International Aviation (CORSIA)(ICAO 2021).

While nearly all the world's airplane operators were required to report CO_2 emissions from 2019, the emission reduction program operates in three phases. A "pilot" phase runs from 2021 to 2023. It is followed by a "voluntary" first phase operating from 2024 to 2026 that includes airplane operators whose national civil aviation authorities decide to participate. From 2027 to 2035, nearly all the world's airplane operators will be required to participate. Baseline emission levels were determined for airplane operators according to their 2019 emissions from international flights. Because of the pandemic in 2020, few if any airplane operators are expected to have compliance obligations during the pilot phase. The degree to which the civil aviation sector recovers its prior passenger loads in 2021 and beyond will determine by how much this policy tool will reduce GHG emissions from this sector.

2.10 Maritime

GHG emissions associated with international maritime transportation in 2012 represented approximately 800 million tCO_2e, or 2.2% of the world's total, not far from the amount emitted by international aviation. Under pressure from climate activists and countries, the International Maritime Organization—the sector's international regulator—set a goal to reduce international maritime emissions 50% by 2050 compared to 2008 levels. This target is all the more ambitious given predictions that international maritime shipping could grow by 50–250% during that period (Lester 2020).

To implement the initial strategy, IMO leaders are "looking at new fuels from renewable and sustainable sources, new methods of propulsion, and new ways of maximizing the efficiency of existing propulsion methods," according to Kitack Lim, president of the IMO (Lester 2020). In 2020, two international banks with a large portfolio of loans to the shipping industry embraced the Poseidon Principles, a voluntary initiative that seeks to hold shippers and their lenders accountable for matching industry climate ambitions to financing decisions and monitoring and reporting. The ultimate goal of the IMO is to achieve full decarbonization of the industry as soon as possible after 2050 (Lester 2020).

2.11 The Limitations of Carbon Trading as a Policy Instrument

A challenge for many cap-and-trade programs established since 2005 has been designing market rules in such a way as to stimulate companies to reduce their emissions rather than simply raising prices to offset the cost of compliance. Regulated companies first sought to reduce emissions associated with inefficient operations. Reducing such emissions also benefited the company by lowering its costs. Structural emission reductions, economists believed, would be economically feasible only when the price of carbon reached €100 per ton. But transformative decarbonization at this scale ran the risk of placing the companies who invested in new methods at a competitive disadvantage with those who did not, particularly if the competitor was located outside the country or region that operated the cap-and-trade system. Designers of cap-and-trade systems understood these risks and generally chose to protect existing industry through such mechanisms as the granting of free carbon allowances.

Policy makers generally have two options for placing a price on carbon emissions. The first is a carbon tax, which is, for example, incorporated into the prices of fossil fuels. Like any tax, the carbon tax is borne by the end consumer, who makes choices between necessity and desire. Only the Nordic countries have a highly dissuasive tax (€130 per ton of CO_2 for Sweden). The second is the price of allowances and offset credits associated with Emissions Trading Systems such as the EU ETS and the California cap-and-trade system.

According to an OECD study (OECD 2018), 90% of emissions have a carbon price of less than €30, and 60% have a price equal to zero. In a study (IC4E 2015) on the factors that have been driving the reduction of CO_2 emissions since 2005 at the facilities concerned, only 0–10% of this decline would be attributable to the price of carbon itself. Fifty percent were due to the deployment of renewable energy, 30% to the economic crisis of 2008–2009, and 10% to 20% to energy efficiency policies.

In the first two decades of the twenty-first century, this regulatory carbon market did not deliver transformative emission reductions because:

- Regulators largely saturated the market with emission allowances to protect large manufacturers from the effects of international competition
- Economic downturns liberated excess emission allowances allocated before companies experienced a reduction in economic activity
- Chronically low-carbon prices did not communicate a realistic price signal to market actors and thus disincentivized low-carbon investment choices.

To reach critical mass, carbon markets are being linked, such as the shared markets of Quebec and California, facilitated by the Western Climate Initiative. However, carbon market participants need to be wary, as political risk is not insignificant. Ontario entered the system in 2017 but exited in June 2018 after the election of a new conservative government.

While cap-and-trade programs in the EU and North America have demonstrated a certain institutional capability—California's has withstood several legal challenges questioning its constitutionality—the larger question is whether they have worked sufficiently well to mitigate GHG emissions and forestall dangerous global warming. Here, the evidence to date is mixed. One lesson learned is that the system of free allocation of allowances must be dramatically scaled back or eliminated. The evidence from programs that have relied upon them is that carbon prices do not rise sufficiently to transmit proper price signals. A turnaround for these shortcomings may have begun in 2021 as EU allowance prices rose to record highs, exceeding €50 per ton for the first time in May of that year and surpassing €60 per ton by September 2021.

2.12 Policy Alternatives to Cap-and-Trade

We have devoted significant attention in this chapter to cap-and-trade as a policy mechanism to address the problem of rising emissions of CO_2 and other greenhouse gases. Cap-and-trade is not, however, the only policy tool available to government decision makers. Two alternatives include carbon taxes and regulation.

2.12.1 Carbon Taxes

Governments impose taxes on goods, labor, property, and capital gains to fund their essential services and to fulfill social contracts with their citizens. They may also impose taxes to dissuade consumption of certain items. Taxes on cigarettes and alcohol, for example, may be motivated in part to encourage abstinence or moderation in use of products deemed to pose a health and safety risk to consumers. Since climate change is arguably a serious threat to health and human welfare, governments are justified in using their taxation powers to raise the cost of GHG pollution and thus encourage the substitution of lower emitting alternatives.

Sweden implemented a carbon tax in 1991 on motor fuel and home heating fuels. Industry was exempt since the start of the EU ETS has been subject to carbon pricing under that cap-and-trade system. Initially, the Swedish tax amounted to €30/t of CO_2, but it had risen to more than €100/t CO_2 by 2020 and was the highest in the world. Sweden reduced other taxes to offset the impact of rising taxes on energy. Overall, carbon emissions in Sweden declined from 1991 to 2018 by approximately 27% (Jonsson et al. 2020).

British Columbia, Canada's Pacific coast province, established a carbon tax on fossil fuels in 2008. The carbon tax is applied to the purchase and use of fossil fuels and covered approximately 70% of provincial greenhouse gas emissions. Initially set at C$10/t$CO_2$e, the carbon tax rate rose to C$40 per t$CO_2$e in 2019. The tax was designed to be "revenue neutral," meaning that other taxes declined to compensate for the rise in carbon taxes. To offset the financial impact of the tax on households, British Columbia's government established an annual Climate Action Tax Credit which by mid-2019 amounted to C$154.50 per adult and C$45.50 per child (British Columbia 2021b).

Canada's provinces have charted varying courses since the start of the Kyoto Protocol's first compliance period to reduce their GHG emissions. From a cap-and-trade program in Quebec to BC's carbon taxes, and to performance standards on heavy emitters in Alberta, the nation's approach has been fragmented. Canada announced its withdrawal from the Kyoto Protocol in 2011. After agreeing to participate in the Paris Agreement, Canada's federal government established a Pan-Canadian Framework on Clean Growth and Climate Change in 2016. This framework allowed provinces to address climate change mitigation through policy means of their choice but created a federal backstop in case provinces failed to meet certain threshold requirements.

The framework imposed a carbon tax of C$10 per tCO_2e on fossil fuels. In 2020 ECCC, the federal ministry announced that the price on carbon emissions for liquid fossil fuels would rise by C$15/$tCO_2e$ per year from 2023, reaching C$170 by 2030. The initial list of backstop provinces where the carbon tax applied included Ontario, New Brunswick, Manitoba, Saskatchewan, Alberta, Yukon, and Nunavut. Ontario and New Brunswick were dropped from the list in 2020 after ECCC determined that their provincial regulations met threshold requirements.

To be effective at changing economic behavior, carbon tax rates need to be high enough to incentivize economic actors to reduce their tax burden by adopting lower GHG emission alternatives to fossil fuel-powered industry, heating, and mobility (Criqui et al. 2009). We discuss in Chap. 8 "The Path to Net Zero" some suggested ways to decarbonize the world's economies.

In 2019, more than 3500 economists, including four former chairs of the US Federal Reserve, 28 Nobel Laureates, and 15 former chairs of the White House Council of Economic Advisors concluded that carbon taxes are the most cost-effective mechanism to decrease emission levels at the speed and scale necessary to meet the world's current climate crisis. They also advocated for "a consistently rising carbon price [to] encourage technological innovation and large-scale infrastructure development" which would "accelerate the diffusion of carbon-efficient goods and services." The statement also advocated for a carbon border adjustment tax to ensure a level playing field for goods manufactured outside the USA that did not have embedded in their prices an equivalent carbon tax (Climate Leadership Council 2019). In 2021, carbon border adjustment taxes were under consideration in the EU.

2.12.2 Regulation

The third common policy instrument of governments is regulation. Laws, and their associated regulations, mandate acceptable behavior by citizens and industries alike. In the USA, the Environmental Protection Agency developed an increasingly thorough set of regulations to control industrial emissions of air pollutants, water discharges, and solid waste disposal. As an example, municipal solid waste facilities are regulated to control wind-blown litter, contamination of groundwater, and odor. They are also required under certain circumstances to install landfill gas collection and combustion systems. According to federal regulations established after the passage in 1995 of the Clean Air Act Amendments, landfills that emit 50 tons per year of non-methane organic compounds (NMOCs) were required to collect and combust their landfill gas (Cornell Law 2006).

Because the threshold in the regulation requiring landfills to collect and combust landfill gas only began at 50 tons of NMOC emissions, most US landfills were not required to control these pollutants and, in the process, also destroy methane which is a powerful GHG. Typical landfill gas is about 50% methane with the remainder comprised mostly of CO_2 and small amounts of the NMOC pollutants that trigger action under the federal regulation. Consequently, the Climate Action Reserve (the Reserve), in version 1 of its Landfill Project Reporting Protocol (2007), determined that landfill gas collection and combustion systems were present in only 9.4–22.3% of those landfills not already required to collect and combust landfill gas (Climate Action Reserve 2019).

The Reserve concluded from this information that installing landfill gas collection and combustion systems at non-regulated (for NMOCs) landfills would be "additional." This paved the way for the Reserve to issue Climate Reserve Tons (CRTs) to project developers investing in landfill gas collection and combustion equipment and destroying the methane contained in the landfill gas. Since 2007, approximately 100 US landfills have earned CRTs by collecting and combusting landfill gas. Another 19 US landfills have earned credits (VCUs) under the CDM ACM0001 methodology.

From a climate perspective, the destroyed tons of methane at municipal solid waste facilities were certainly "real." Absent the projects, methane would have escaped the landfills as fugitive emissions. The projects passed the additionality test because existing regulations did not require the landfills to collect and destroy landfill gas. This was true in all 50 US states until December 2013 when California's revised regulations *required* municipal solid waste landfills in that state to collect and destroy landfill gas.

As we will see in Chap. 4, "Reducing Emissions," human ingenuity has found many ways to make real emission reductions. The question is why have governments not required greater control of GHG emissions at the source and enacted regulations to ensure, like California and British Columbia did with their municipal solid waste landfills, that fugitive, process, and direct emissions from industrial activity were not controlled to the extent possible using best available technology? The most plausible answer to that question was that policy makers in most jurisdictions simply lacked the political will—or votes—to take action to combat the climate change crisis.

2.13 The Importance of the Price Signal

One explanation for the worldwide increase in GHG emissions since 1990 is the lack of an adequate price signal to change behaviors. Voluntary carbon credits for some project types have at the most sold for about $5/ton, but the long-term average price per ton for voluntary carbon credits over the period 2007–2020 has been closer to $1 to $2 per ton and sometimes substantially lower. Emission allowances under mandatory cap-and-trade regulations have traded at higher prices, but insufficiently high to incentivize the regulated community to decarbonize heavy emitting industries.

Until at least 2020, the voluntary market served as a reservoir of low-cost carbon credits that companies could tap to "offset" their emissions of GHGs. Companies in consumer-facing sectors purchased credits to substantiate claims of "carbon neutrality" and show their support for climate action. By 2020, more than 1000 firms had pledged to align their emissions with the Paris Agreement, which meant reducing their emissions to zero by 2050 (Turner and Grocott 2020). As complete decarbonization of operations remains a distant goal for most industries, achieving carbon neutrality or "net-zero" emissions are based on the purchase of carbon credits to balance credit and debit entries in a firm's GHG inventory.

In 2021, corporations demonstrated an increased willingness to acknowledge the urgency of the climate crisis. This recognition is welcome, and all efforts made by business, finance, and industry to decarbonize economies are beneficial. Nonetheless, purchasing ultra-low-cost voluntary carbon credits does not address the problem, because the purchased offset credits' reductions do nothing to reduce hard-to-abate emissions. The additionality of these credits is questionable, and rather than contributing to decarbonization, organizations purchasing low-quality carbon credits may only be delaying taking the necessary actions to decarbonize their processes and products.

In late 2020, the world was awash in unsold carbon credits. According to a Trove Research report (Turner and Grocott 2020):

1. "There are some 580 Mt[7]CO_2e of surplus credits currently in all the main voluntary carbon offset registries (cumulative issued credits less the volume retired). This represents more than seven times the current demand for carbon offsetting of around 80 MtCO_2e in 2020.
2. "The surplus is still increasing. In 2020, we estimate that 200 Mt of credits will be issued but only 80 Mt retired, adding 120 MtCO_2e to the surplus.
3. "The potential for projects to back-issue even more credits is even greater. We estimate that the total surplus in the voluntary carbon registries could amount to 1200 Mt. This would represent × 15 the current annual demand.
4. "The potential volume of accumulated carbon credits in the CDM is even greater. Many of these credits have more dubious claims of environmental additionality. If all registered CDM projects elected to have their carbon credits verified for the previous decade, they would produce a total additional volume of nearly 7000

---

[7] Mt = million tons.

$MtCO_2e$. This would represent × 80 current annual demand. If allowed into the voluntary market, these CDM credits would effectively make the voluntary market redundant as a mechanism to reduce global carbon emissions."

The problems that this situation raises should be obvious. Corporations seeking to do the "right thing" by offsetting GHG emissions can purchase voluntary carbon market credits at very low cost—far lower than their own cost of carbon emission abatement. However, some credits readily available for purchase have dubious claims to additionality. "Offsetting" with credits of this kind in no way tackles the difficult job of decarbonizing the hard-to-abate industries that contribute the most to the global problem of excess carbon emissions. Much higher prices for voluntary carbon credits are needed to ensure the environmental integrity of the offset claims that corporations are making.

Chap. 8 "The Path to Net Zero" describes some promising technological innovations that may contribute to the decarbonization of economies that the world desperately needs.

Questions for Readers

1. What do you think the effect was of the Kyoto Protocol coming into force without the participation of the USA and China?
2. What are the inherent limitations of cap-and-trade programs that do not include as participants all industrialized economies?
3. What relevance do declines in per capita GHG emissions have on the ability of the world to keep global warming to between 1.5 and 2 °C?
4. What do you believe the relative advantages and disadvantages are between the imposition of carbon taxes and the establishment of cap-and-trade systems?
5. What is the importance of setting a "price signal" for emissions of greenhouse gases and the level of that signal?

References

3Degrees (2017) First RGGI carbon offset project launched by 3Degrees. https://3degreesinc.com/news/rggi-carbon-offset-first-3degrees/. Accessed 23 May 2021

Alberta (2018) Carbon competitiveness incentive regulation fact sheet. https://www.alberta.ca/assets/documents/cci-fact-sheet.pdf. Accessed 26 Dec 2020

British Columbia (2020) Greenhouse gas reduction (Renewable and low carbon fuel requirements) act, renewable and low carbon fuel requirements regulation. https://www.bclaws.gov.bc.ca/civix/document/id/crbc/crbc/394_2008. Accessed 16 Dec 2020

British Columbia (2021a) Greenhouse gas industrial reporting and control Act. https://www.bcl aws.gov.bc.ca/civix/document/id/lc/statreg/14029_01#section8. Accessed 24 May 2021

British Columbia (2021b) British Columbia's carbon tax. https://www2.gov.bc.ca/gov/content/env ironment/climate-change/clean-economy/carbon-tax. Accessed 12 Dec 2020

Broughton E (2005) The Bhopal disaster and its aftermath: a review. Environ Health. https://doi. org/10.1186/1476-069X-4-6. Accessed 24 Dec 2020

C2ES (2021) California's cap and trade. https://www.c2es.org/content/california-cap-and-trade/. Accessed 15 Dec 2020

CARB (2015) Staff report: initial statement of reasons for proposed rulemaking. https://ww3.arb. ca.gov/regact/2015/lcfs2015/lcfs15isor.pdf. Accessed 15 Dec 2020

CARB (2017) 2017 scoping plan documents. https://ww2.arb.ca.gov/our-work/programs/ab-32-cli mate-change-scoping-plan/2017-scoping-plan-documents. Accessed 23 May 2021

CARB (2020) 2020 Mobile source strategy. https://ww2.arb.ca.gov/resources/documents/2020-mobile-source-strategy. Accessed 19 Dec 2020

CARB (2021) Current California GHG emission inventory data. https://ww2.arb.ca.gov/ghg-invent ory-data. Accessed 23 May 2021

Carbon Market Watch (2013) Doha on AAUs: the future of the Phantom menace (Newsletter #2). https://carbonmarketwatch.org/2013/03/04/doha-on-aaus-the-future-of-the-pha ntom-menace/. Accessed 10 Feb 2021

CBC (2018) Ontario government officially kills cap-and-trade climate plan. https://www.cbc.ca/ news/canada/toronto/ontario-officially-ends-cap-and-trade-1.4885872. Accessed 26 Dec 2020

CDM (2021) Clean development mechanism: CDM projects. https://cdm.unfccc.int/index.html. Accessed 6 June 2021

CleanBC (2020) 2020 Climate change accountability report. https://www2.gov.bc.ca/assets/gov/ environment/climate-change/action/cleanbc/2020_climate_change_accountability_report.pdf. Accessed 26 Dec 2020

Climate Action Reserve (2019) U.S. Landfill protocol development. http://www.climateactionre serve.org/how/protocols/us-landfill/dev/. Accessed 15 Jan 2021

Climate Leadership Council (2019) Economists' statement on carbon dividends. https://clcouncil. org/economists-statement/. Accessed 27 Dec 2020

Cornell Law (2006) 40 CFR § 60.752—standards for air emissions from municipal solid waste landfills. https://www.law.cornell.edu/cfr/text/40/60.752. Accessed 15 Dec 2020

Criqui P et al (2009) States and carbon. https://www.researchgate.net/publication/47281181_Pat rick_Criqui_Benoit_Faraco_Alain_Grandjean_2009_Les_Etats_et_le_carbone_Presses_Univer sitaires_de_France_France_192_p. Accessed 5 Feb 2021

ECCC (2021) What is the clean fuel standard? https://www.canada.ca/en/environment-climate-cha nge/services/managing-pollution/energy-production/fuel-regulations/clean-fuel-standard/about. html. Accessed 18 June 2021

European Commission (2021a) EU emissions trading system (EU ETS) https://ec.europa.eu/clima/ policies/ets_en. Accessed 18 June 2021

European Commission (2021b) Communication from the commission to the European parliament, the council, the European economic and social committee and the committee of the regions, commission work programme 2021 A union of vitality in a world of fragility available at https://eur-lex.europa.eu/resource.html?uri=cellar%3A91ce5c0f-12b6-11eb-9a54-01aa75ed7 1a1.0001.02/DOC_1&format=PDF. Accessed 24 May 2021

European Council (2007) Presidency conclusions Brussels European council 21/22 June 2007 11177/1/07 REV 1 p 10 available at https://www.consilium.europa.eu/ueDocs/cms_Data/docs/ pressData/en/ec/94932.pdf. Accessed 24 May 2021

European Council (2020) Climate change: what the EU is doing. https://www.consilium.europa.eu/ en/policies/climate-change/. Accessed 30 Dec 2020

Fabiano B et al (2017) A perspective on Seveso accident based on cause-consequences analysis by three different methods. In: Journal of loss prevention in the process industries. Science direct. https://www.sciencedirect.com/science/article/abs/pii/S0950423017300864. Accessed 24 Dec 2020

Fenech A et al (2003) Natural capital in ecology and economics: an overview. Environ Monit Assess. https://doi.org/10.1023/A:1024046400185. Accessed 24 Dec 2020

Green Climate Fund (2021) About GCF. https://www.greenclimate.fund/about. Accessed 7 Feb 2021

Guardian (2011) Canada pulls out of Kyoto protocol. https://www.theguardian.com/environment/2011/dec/13/canada-pulls-out-kyoto-protocol. Accessed 12 Dec 2020

ICAO (2021) Carbon Offsetting and Reduction Scheme for International Aviation (CORSIA). https://www.icao.int/environmental-protection/CORSIA/Pages/default.aspx. Accessed 6 June 2021

IETA (2015) Alberta's climate change program. https://www.ieta.org/page-18192/3501548. Accessed 26 Dec 2020

Jain P et al (2017) Did we learn about risk control since Seveso? Yes, we surely did, but is it enough? An historical brief and problem analysis. In: Journal of loss prevention in the process industries. Science direct. https://www.sciencedirect.com/science/article/abs/pii/S0950423016302674. Accessed 24 Dec 2020

Jonsson S et al (2020) Looking back on 30 years of carbon taxes in Sweden. https://taxfoundation.org/sweden-carbon-tax-revenue-greenhouse-gas-emissions/. Accessed 12 Dec 2020

Kerr T (2007) "Voluntary climate change efforts". In Gerrard M, (ed) Global climate change and U.S. Law, American Bar Association, pp 614–615

Lester A (2020) Shipping—the trillion-dollar challenge. https://www.environmental-finance.com/content/analysis/shipping-the-trillion-dollar-challenge.html. Accessed 27 Dec 2020

McDowell T (2017) The case for cap-and-trade: California's battle for market-based environmentalism. https://harvardelr.com/2017/11/03/the-case-for-cap-and-trade-californias-battle-for-market-based-environmentalism/. Accessed 15 Dec 2020

McInnes C (2008) Dion's carbon tax easier to attack than explain. https://vancouversun.com/news/staff-blogs/dions-carbon-tax-easier-to-attack-than-explain. Accessed 16 Dec 2020

Miller M, Batchelor T (2012) Information paper on feedstock uses of ozone-depleting substances, produced for the European commission, December 2012. https://ec.europa.eu/clima/sites/clima/files/ozone/docs/feedstock_en.pdf. Accessed 2 Jan 2021

Millington D et al (2020) "The economic effectiveness of different carbon pricing options to reduce carbon dioxide emissions. "Study No. 189. Calgary, AB: Canadian energy research institute. https://ceri.ca/assets/files/Study_189_Full_Report.pdf. Accessed 1 March 2021

Natural Resources Canada (2020) Crude oil facts. https://www.nrcan.gc.ca/science-data/data-analysis/energy-data-analysis/energy-facts/crude-oil-facts/20064. Accessed 26 Dec 2020

Ontario (2019) O. Reg. 241/19: greenhouse gas emissions performance standards. https://www.ontario.ca/laws/regulation/r19241. Accessed 26 Dec 2020

Ontario (2020) Amendments to transition Ontario industrial facilities from the federal output-based pricing system to Ontario's emissions performance standards program. https://ero.ontario.ca/notice/019-2813. Accessed 26 Dec 2020

Ontario Ministry of Finance (2021) Ontario fact sheet https://www.fin.gov.on.ca/en/economy/ecupdates/factsheet.html. Accessed 31 May 2021

Oregon Secretary of State (2019) Oregon clean fuels program. https://secure.sos.state.or.us/oard/displayDivisionRules.action;JSESSIONID_OARD=mO-BjHDCoRovT20WM9wbqRj9_lm2xm0TewVyJYkDm06s_sobz7n3!-1835049044?selectedDivision=1560. Accessed 15 Dec 2020

RGGI (2021a) A brief history of RGGI. https://www.rggi.org/program-overview-and-design/design-archive. Accessed 23 May 2021

RGGI (2021b) Supply and bid statistics. https://www.rggi.org/Auctions/Auction-Results/Supply-Bid. Accessed 23 May 2021

Slater H et al (2021) China's national carbon market is about to launch. https://chinadialogue.net/en/climate/chinas-national-carbon-market-is-about-to-launch/. Accessed 15 Feb 2021

Turner G, Grocott H (2020) The global voluntary carbon market. Dealing with the problem of historic credits. 4 December 2020. Trove research. https://trove-research.com/research-and-insight/the-global-voluntary-carbon-market-dealing-with-the-problem-of-historic-credits-dec-2020/. Accessed 27 May 2021

UNFCCC (2015) Article 14. In: Paris agreement. https://unfccc.int/sites/default/files/english_paris_agreement.pdf. Accessed 27 Dec 2020

UNFCCC (2016) NDC Interim Registry. https://unfccc.int/news/ndc-interim-registry. Accessed 31 December 2020

United Nations (2004) Meetings coverage and press releases. https://www.un.org/press/en/2004/envdev793.doc.htm. Accessed 31 Dec 2020

United Nations (2021) Sustainable development goals https://sdgs.un.org/goals. Accessed 24 May 2021

Chapter 3
Counting Carbon

3.1 The Context for Counting Carbon

Tools exist to measure an organization's emissions of greenhouse gases—the first step toward developing and executing strategies for decarbonization. Fortunately, many of these strategies have the side benefit of reducing costs for purchased energy, materials, and services. The management approaches to achieve greenhouse gas (GHG) emission reductions are not new, but they do need to be deployed more extensively. ISO 14001, Environmental management systems—Requirements with guidance for use, offers each organization a structured approach for considering the environmental aspects and impacts of its activities, products, and services. First published in 1996, the 2015 third edition of ISO 14001 provides additional clarity about an organization's obligation to consider both its internal and external contexts when designing and implementing its environmental management system (EMS). The external context includes such environmental condition indicators as global warming and related climate change impacts (ISO 2015).

Some organizations also have found useful ISO 50001, Energy management systems—Requirements with guidance for use. This standard, first published in 2011 and revised in 2018, adapted the ISO 14001 environmental management approach to the specific purpose of managing an organization's consumption of energy with the goal of achieving performance improvements. Reducing energy consumption provides many benefits, starting with the reduction in energy costs at the organizational level. But the benefits do not stop there. When many consumers reduce energy consumption, an electric utility can serve its stable or increasing customer base without adding new generating facilities. In some cases, more polluting generating stations, such as those fired by coal, can be shut down as electricity loads are reduced through the efficient use of energy and increasing supply from renewable energy sources (ISO 2018a).

Improving energy efficiency is one reason why organizations should establish a GHG inventory. According to the principle "what gets measured, gets managed," an inventory of GHG emissions at the organizational level defines the organization's

© The Author(s), under exclusive license to Springer Nature Switzerland AG 2021
J. C. Shideler and J. Hetzel, *Introduction to Climate Change Management*,
Springer Climate, https://doi.org/10.1007/978-3-030-87918-1_3

carbon footprint and serves as the starting point for setting and meeting GHG emission reduction goals. One tool for designing an organizational GHG inventory is ISO 14064 Part 1, Specification with guidance at the organizational level for quantification and reporting of greenhouse gas emissions and removals. ISO 14064 Part 1, revised in 2018, guides organizations through the process of defining the scope of their GHG emissions reporting, quantifying emissions to and removals of carbon from the atmosphere, managing the quality of data and information, and reporting aggregated and consolidated GHG emissions. Finally, it describes the role an organization plays in preparing for verification of reported emissions and removals (ISO 2018b).

ISO 14064 Part 1 describes an approach to accounting of GHG emissions and removals that resembles in many ways an organization's statement of financial information. Like a statement of financial results, an organization reporting GHG emissions and removals aggregates quantitative information from individual facilities and sources into a consolidated statement that reports on total emissions and removals at the organizational level. In GHG accounting, emissions are similar to "debits" in a financial accounting system, while removals are similar to "credits." However, unlike financial accounting, which rewards organizations for improving their profits and adding to their capital, emissions, which in nearly every organization far exceed removals, are a measure of negative impact on the global warming of the planet. Despite this upside-down perspective on the numbers, in many other respects accounting for GHG emissions and removals is remarkably like accounting for financial results.

3.2 Selecting Reporting Criteria

The choices facing an organization wishing to report its GHG emissions are many. One of them is deciding among various criteria that offer standardized methods for monitoring and reporting emissions. These rules prescribe what to report and how. Organizations reporting to a regulatory authority have their criteria specified for them: They report in accordance with the rules of the mandatory reporting program. Organizations reporting for their own benefit have options to choose from. Some organizations are obliged to use different methods for reporting for different purposes—regulatory requirements in one or more jurisdictions, and voluntary reporting for the corporation as a whole.

3.2.1 The GHG Protocol

By far, the most widely adopted standard and guidance document for corporate reporting of GHGs was published by the World Resources Institute and the World Business Council for Sustainable Development (WRI/WBCSD). Documents in this series include the GHG Protocol Corporate Accounting and Reporting Standard (revised edition 2004, hereafter the "GHG Protocol" [2004]), the GHG Protocol Scope 2 Guidance, an amendment to the GHG Protocol Corporate Standard (2015), and the GHG Protocol Corporate Value Chain (Scope 3) Accounting and Reporting Standard (2011). The GHG Protocol categorizes emissions by "scope," where scope 1 refers to gases emitted directly by the organization through the combustion of fuels, industrial processes, or leaks. Emissions associated with imported electricity, heat, and cooling are categorized as scope 2, and upstream and downstream emissions associated with purchased goods and services, or with consumer end uses, are categorized as scope 3.

3.2.2 ISO 14064-1

The International Organization for Standardization (ISO) published the first edition of ISO 14064 Part 1 in 2006. The first and second editions are generally consistent with the GHG Protocol standards. ISO standards are widely used in industry around the world and are sometimes incorporated by countries in national regulations. The World Trade Organization considers ISO standards to represent an international consensus and therefore recognizes them as not constituting a technical barrier to trade when their use is mandated by governments.

3.2.3 Other Criteria

Many other criteria exist. These criteria include voluntary reporting programs, such as the CDP, the Climate Registry, and Airport Carbon Accreditation. Voluntary programs often reference either the GHG Protocol or the ISO standard. Regulatory reporting programs typically are inspired by the same principles and methods found in the GHG Protocol or the ISO standard. Many mandatory programs exist at the national or state and provincial government levels in North America. Mandatory reporting programs tend to focus on direct emissions only and may have very detailed requirements describing how regulated industries must quantify and report these emissions.

3.3 Organizational Boundaries of the Inventory

After selecting criteria, the GHG inventory manager next determines the organizational boundaries of GHG reporting. The word "organization" provides some flexibility as ISO 14064 Part 1 defines it as "company, corporation, firm, enterprise, authority or institution, or part or combination thereof, whether incorporated or not, public or private, that has its own functions and administration" (ISO 2018b). Thus, an organization may set its organizational boundary for reporting purposes at a highly consolidated level, such as at the level of a multinational corporation, or at a more disaggregated level, such as at the level of an individual university campus in a multi-university system. At whatever level it is defined, the organization should have its own functions and administration and thereby be capable of managing and reporting its GHG inventory. In some circumstances, an individual "facility" may be designated as a reporting unit. For example, some governmental programs require reporting at the industrial facility level regardless of the dependence their personnel may have on centralized management for greenhouse gas data collection and reporting.

The organizational boundary helps determine the scope of emissions or removals to be reported. According to the principle of completeness, all emissions and removals within the boundaries of the organization should be included. Exceptions may be made, however, for truly minor sources that the organization does not deem to be "relevant." Inventory reporting typically begins at the facility level, where emissions from individual sources are aggregated in direct and indirect categories. When an organizational boundary comprises more than one facility, emissions and removals are consolidated at the higher organizational level. ISO 14064 Part 1 gives organizations the choice to consolidate emissions and removals using the criterion of control, either operational or financial, or by equity share.

Operational control exists when an organization has the full authority to introduce and implement operating policies at the operational level (ISO 2018b). This is the most common form of consolidation of emissions data in organizations. However, organizations may also consolidate emissions data under financial control. An organization exercises financial control when it can direct the financial and operating policies of an operation with a view to gain economic benefits from its activities. In most circumstances, there are few differences between these two approaches to consolidation. Variances typically concern subsidiaries and other related parties. Larger and more complex organizations may wish to follow the same consolidation approach for GHG emissions reporting as they do for reporting financial results. This means that if the assets and earnings of a subsidiary are consolidated for financial reporting purposes according to the financial control method, the organization should do the same for its GHG inventory reporting.

Another approach to consolidation of emissions reporting is equity share. ISO 14064 Part 1 describes this approach as reporting the percentage of economic interest in, or benefit derived from, a facility. Oil and gas companies and electric utilities may use this approach when the ownership of an asset is governed by a production sharing agreement. Under equity share, each partial owner reports only the pro rata share of emissions that is associated with its percentage of ownership, regardless of which company actually operates the asset. The equity share approach may be preferred when it mirrors the way an organization reports its financial results. Some organizations choose to report by one of the control approaches *and* equity share, in order to contrast the different results of both approaches.

Once the organizational boundary and method of consolidation of results have been established, a reporting boundary can be defined that includes all GHG emissions and removals that the organization should report. ISO 14064 Part 1 distinguishes two categories of emissions and removals. Direct emissions and removals are those that occur due to processes and activities controlled by the organization. Indirect emissions, on the other hand, result from the actions of organizations outside the organizational boundary of the reporting entity. Indirect emissions, such as the consumption of electricity from the grid and imported steam, are reported by organizations that use them because the emissions they represent would not have occurred absent the organization's demand for the purchased energy or other product or service. This does not constitute "double counting" of emissions because the energy indirect emissions are reported separately and clearly labeled as "indirect." Readers of the inventory report therefore know that another organization was responsible for generating them.

Direct emissions are usually grouped in the following categories:

- Stationary combustion
- Mobil combustion
- Process
- Fugitive
- Land use, land use change, and forestry.

GHG inventory managers whose task is to identify individual sources of emissions will find that all direct emissions may be traced to one of these groups.

Removals occur when carbon is sequestered through biological processes such as photosynthesis and by deposition and accumulation in soil. Removals are prominently associated with the agriculture and forestry sectors. A prime example of a source for removals is a tropical rain forest such as the Amazon, which is sometimes referred to as the lungs of planet Earth. Incidental removals, which occur when organizations maintain properties with trees used in landscaping, typically are not reported as the amount of carbon sequestered is small and they fail the test of "relevance" to the inventory as a whole.

Carbon capture and storage (CCS) is a technology that holds promise for slowing the release of GHGs into the atmosphere. At present, CCS is difficult to engineer and expensive to build and operate, so its use outside the oil and gas industry has been limited to a few pilot projects in the electricity generation sector. The technology has found greater success in the oil and gas industry which has reinjected CO_2 removed from gas associated with oil production into producing formations to enhance further oil recovery. Limiting CO_2 emissions to the atmosphere by pumping them into geologic formations is desirable, and where it occurs, it should be accounted for. However, where CO_2 is injected into geologic formations, these same formations should be monitored for leaks, and any fugitive releases reported as direct emissions.

CO_2 emissions from the combustion of biomass are quantified and reported separately. Where biomass is used for energy production, and where the biomass derives from annual crops, "IPCC Guidelines assume that biomass carbon stock lost through harvest and mortality equal biomass carbon stock gained through regrowth in that same year and so there are no net CO_2 emissions or removals from biomass carbon stock changes"[1] (IPCC 2021). ISO 14064 Part 1 requires CO_2 emissions from the combustion of biomass to be reported as a separate line item. When burning biomass for heat generation, the combustion by-products methane (CH_4) and nitrous oxide (N_2O) are considered anthropogenic and are reported as a direct emission of the organization.

Where organizations have identified fuel combustion to be a significant inventory component, the life cycle emissions associated with production of those fuels may be considered relevant and reported as indirect emissions. For fossil fuels, this includes emissions from petroleum and natural gas production, processing, and transportation; for biofuels, this includes emissions from feedstock cultivation and harvest, land use change, processing, and transportation (ISO 2018b).

Indirect emissions include those associated with energy imported into the organization as well as the GHG emissions associated with products and services consumed or used by the organization. The second edition of ISO 14064 Part 1 groups all indirect emissions into a single "indirect" category, in contrast to the first edition which subdivided them into "energy indirect" and "other indirect" categories. Another departure from the first edition is the requirement that the organization "applies a process to determine which indirect emissions to include in its GHG inventory." Moreover, the organization "shall identify and evaluate its indirect GHG emissions to select the significant ones. The organization shall quantify and report these significant emissions. Exclusions of significant indirect emissions shall be justified" (ISO 14064: 2018b, 5.2.3).

This treatment of indirect emissions is consistent with the third edition of ISO 14001, which promotes the use of a life cycle perspective to environmental management (ISO 2015, 6.1.2 and 8.1). An organization thus is encouraged to consider the global warming impact of both its own activities, products, and services and also the global warming impact of those activities, products, and services that it incorporates from suppliers and that it passes down to customers. The resulting "supply

---

[1] Accounting for biomass CO_2 emissions can be controversial. The IPCC Task Force on National Greenhouse Gas Inventories (2021) "do[es] not automatically consider biomass used for energy as "carbon neutral," even if the biomass is thought to be produced sustainably, because:

(1)   "In any time period there may be CO_2 emissions and removals due to the harvesting and regrowth of bioenergy crops;

(2)   "Land use changes caused by biomass production can also result in significant GHG fluxes; and

(3)   "There may also be significant additional emissions which are estimated and reported in the sectors where they occur e.g.:

    a.   "From the processing and transportation, etc., of the biomass

    b.   "Direct methane and nitrous oxide emissions from the biomass combustion

    c.   "From the production and use of fertilizers and liming if either is used in cultivation of the biomass.

"For example, direct methane and nitrous oxide emissions from biomass combustion for energy use are reported in the energy sector" (Question 2–10).

chain" accounting of the accumulated global warming impacts may be expressed as a "carbon footprint" for individual products using quantification and reporting standards such as ISO 14067 (ISO 2018c).

3.4 Quantification of GHG Emissions and Removals; Emissions from Sinks; and Fluxes in Reservoirs

If an organization quantifies GHG removals, it will identify and document the corresponding GHG sinks such as forests, grasslands, and agricultural soil. It should be noted that inventory reporting should quantify only the increases in removals from a prior period and not the total stocks of carbon in the sink. Net decreases in removals may also occur in any sink category. For example, harvesting forest timber or the occurrence of a forest fire reduces carbon stocks. These reductions should be reported as "emissions" with respect to the identified sink.

A reservoir that stores greenhouse gases is neither a source nor a sink under normal operating conditions. Natural gas storage reservoirs are common features in industrialized economies and serve the important role of regulating the balance between demand and supply. As long as gas is added to and removed from the reservoir under controlled conditions, no emissions are reported in GHG inventory accounts. However, reservoirs should be monitored for leaks and the quantity of resulting GHGs released to the atmosphere counted as an "emission."[2]

Quantification methodologies rely on either measurement or calculation, or a combination of both. Direct measurement of GHGs occurs using continuous or intermittent gas analyzers in the oil and gas, power production, and chemical industry sectors, among others. (See Chap. 5 for more detail on monitoring technology.) For many sources, however, emissions are calculated using a combination of activity data and emission or removal factors. Activity data provide a "quantitative measure of activity that results in a GHG emission or GHG removal" (ISO 2018b). For example, the amount of fuel combusted in a road transportation vehicle constitutes "activity data." The GHG emissions associated with the combustion of that fuel can be calculated by the application of an emission factor that is based upon stoichiometry (the calculation of relative quantities of reactants and products in chemical reactions). In the case of fuel combustion from road transportation, the emissions factor provided by one authority for diesel is 2681 gCO_2 per liter of diesel combusted (Climate Registry 2020). If a company uses 100,000 L of diesel in a year, its emissions of CO_2 are 268,100,000 g or 268.1 metric tons.[3]

The use of emission factors greatly simplifies the quantification of GHG emissions. However, the precision of emissions factors varies. Pipeline quality natural

---

[2] A major uncontrolled leak of natural gas from the Porter Ranch gas storage facility in Southern California released 30 tons of methane per hour for months in late 2015 and early 2016 (Barboza 2016).

[3] "Metric tons" are also known as "megagrams" in the *Système International* (SI) of measurement.

gas may vary by up to 2% in heating value, which means that the use of a "default" emissions factor may understate or overstate actual emissions from the use of natural gas. In cases where greater precision is desired, the reporting entity can obtain the actual heating value associated with delivered natural gas and calculate emissions based on the energy content of the delivered gas. Most providers of natural gas base their invoices on the actual heating value delivered to customers rather than upon volume, so the equation for calculating emissions from this fuel is relatively simple.

Many industrialized countries require GHG emissions reporting from facilities located within their jurisdictions. Regulated entities typically include cement plants, manufacturers with large amounts of stationery combustion, refineries, chemical plants, pulp and paper mills, steel plants, and others.

The regulatory requirement to quantify and report GHG emissions by itself has only a small impact on industrial operators. Mandatory reporting alone does not constitute a policy that will drive down emissions and lead to a transition to a low-carbon strategy. To properly mobilize management to embrace a low-carbon strategy, other policy levers are required as we discuss in more detail in Chap. 8.

3.4.1   Fugitive Emissions

Knowledge of industrial, technological, or waste management practices leads to the identification of places where natural or engineered gases may leak into the environment. With respect to refrigeration or electrical switchgear equipment, the best method of accounting is to analyze the difference between the specified capacity of industrial process equipment and the quantities of gases needed to restore it to full capacity. Such analyses may point to losses of HFC refrigerants associated with cooling systems or of SF_6 used to insulate high-voltage electricity transmission equipment. Both of these are sources of high-GWP greenhouse gases. Monitored activity data are difficult or impossible to obtain for these source types. Instead, a mass balance approach is used when equipment is recharged with refrigerants or insulting gases, or default leakage rates may be applied based on engineering estimates. Fugitive emissions are also common in oil and gas production where high-pressure bleed valves that operate using pressurized natural gas from a well are partially leaked to the atmosphere to actuate a device and from oil storage tank pressure relief valves. Like oil fields, pipelines are also a source of fugitive emissions.

Fugitive emissions of GHGs often involve CH_4, N_2O, hydrofluorocarbons (HFCs), perfluorocarbons (PFCs), and sulfur hexafluoride (SF_6). The global warming potentials of these gases, when normalized to equivalent amounts of CO_2, are expressed in tons of carbon dioxide equivalents (CO_2e). We discuss global warming potentials in more detail in Chap. 1.

3.4.2 GHG Removals

In the case of a forest sink, the amount of carbon sequestered in a year is typically estimated through the use of a model. Direct measurement of carbon sequestered in a tree is not feasible, so foresters typically use a growth and yield model to estimate the amount of carbon sequestered in a given time period. The process is complex, and the results of the model are checked periodically (such as every five or six years) by measuring the diameter of a sample of standing trees in a forest and sometimes by counting the amount of woody biomass that is found on the forest floor. Growth and yield models are developed for different tree species by national forestry authorities or academic or industry experts. Foresters use the same modeling methods to estimate the amount of merchantable standing timber in a forest for asset evaluation or sales purposes.

3.4.3 Data Sources for GHG Quantification

The most accurate data for quantifying GHG emissions are localized and site-specific. These data are generally known as "primary data." When primary data are not available, secondary data from the literature or recognized databases may be used.[4] In reporting GHG data and information, organizations should explain their approach to quantifying emissions. When modifying previously used quantification methods, organizations should disclose changes to inventory practices in subsequent reports.[5]

Many GHG programs stipulate the quantification methods that organizations shall use in reporting GHG emissions. Direct measurement is preferred but not always practicable. More often, GHG emissions and removals are modeled, with varying degrees of accuracy and uncertainty. The most common form of modeling is based upon "activity data"—such as the volume or mass of fuel consumed in a combustion device—and an "emissions factor." This type of quantification method yields very accurate results for GHG emissions from liquid and gaseous fuels, and somewhat less accurate results for combustion using solid fuels, particularly wood. An important factor to know when burning wood as a solid fuel is its moisture content.

Biogenic CO_2 emissions are treated separately from CO_2 emissions from anthropogenic sources. CO_2 from the use of renewable ethanol in fuels or biogas produced in landfills or dairy digesters are examples of biogenic CO_2 emissions that an organization could report separately from fossil fuel combustion. Note, however, that the combustion by-products (CH_4 and N_2O) of biogenic fuels are reported as anthropogenic sources.

---

[4] ISO 14033 (2019) uses different terms to describe quantitative environmental data.

[5] More in-depth information about quantitative information may be found in ISO 14033 (2019), *Environmental management—Quantitative environmental information—Guidelines and examples.*

3.4.4   Indirect Emissions from Imported Energy

Indirect GHG emissions from the generation of imported[6] electricity, steam, heat, and cooling are reported by each organization as an indicator of that organization's energy intensity. Imported electricity results in indirect emissions to the organization that buys it because the energy producer accounts for the direct emissions. If an organization generates its own electricity, those emissions are not indirect but direct and are reported just like any other emissions from the combustion of fuel. Emission factors for imported electricity should account for fossil fuel consumed in the generation of electricity. Related but different categories of indirect emissions account for transmission and distribution losses, and any emissions associated with plant construction that are amortized in accordance with established financial practice. Indirect emissions should be quantified and reported separately from the organization's direct emissions (ISO 2018b, 5.2.4).

An organization reporting indirect emissions should seek emission factors from its suppliers. Where this is not possible, or where a supplier generates electricity from a mix of generating sources, organizations should use a subnational average emissions factor as first choice or a national average as second choice.

Renewable electricity certificates or other contractual offset credits linked to renewable energy generation may be reported separately but are not subtracted from the organization's emissions.

Some GHG programs require the use of a dual reporting approach including both "location-based" and "market-based" methods for quantifying indirect emissions from the use of imported electricity (Greenhouse Gas Protocol 2015). The market-based approach takes into account an organization's purchase of renewable energy certificates (RECs) representing renewable energy generation. Reporting using the "market-based" method is intended to incentivize the use of alternative low-carbon energy resources at the organizational level.

The use of default emission factors greatly simplifies the calculation of GHG emissions. Emission factors should be chosen to reflect local conditions most closely. When local emission factors are not available, GHG inventory managers should select from subnational emissions factors, national emissions factors, or international emissions factors in descending order of preference. In the USA, the Environmental Protection Agency (US EPA) has divided the country into regional grids for electricity generation. Annually, the EPA publishes emissions factors for each grid. The emissions factors vary considerably from one region to another depending on the proportion of renewable sources, nuclear power plants, and coal-fired power plants in each grid.

---

[6] "Imported" means that the electricity or other energy-containing product was generated outside the operational boundary of the organization and is delivered to the organization. Typically, this happens when a power plant operator sells electricity to its customers.

3.4.5 Indirect Emissions Other Than Imported Energy

"Scope 3" emissions[7] are those that result from the company's upstream and down-stream activities, other than emissions included in scope 1 and scope 2. For many companies, scope 3 is the largest category of emissions in the organizational inventory. Typically, the largest source of scope 3 emissions derives from an organization's supply chain which may stretch around the world. Supply chains are made up of individual companies that provide factories with raw materials or produce parts that ultimately are combined at a final assembly plant to create a finished product. Supply chain transportation-related emissions are also included in this category.

Policies that only focus on reduction of scope 1 and scope 2 emissions miss a large proportion of GHGs that are embodied in final products. For this reason, the second edition of ISO 14064 Part 1 abandoned the three-category approach and requires users to report all significant sources of indirect GHG emissions (ISO 2018b). The evaluation of scope 3 is the subject of different methodologies, but it is mainly a question of using one of the following two methods:

- Inventory method using activity data and emission factors
- Life cycle assessment approaches resulting in carbon footprints of product which may use either site-specific activity data and emissions factors or information derived from databases that represent industry average emissions data.

Carbon accounting methods make it possible to reasonably quantify GHG emissions related to business-to-business transactions, even though the precision of such estimations is usually lower than that for direct emissions and emissions associated with purchased energy. The importance of reporting upstream and downstream emissions to obtain a fuller picture of an organization's contribution to global warming prompted WRI/WBCSD to publish its GHG Protocol Corporate Value Chain (Scope 3) Accounting and Reporting Standard in 2011.

The GHG Protocol followed its standard with a comprehensive guidance document for the calculation of scope 3 emissions (Greenhouse Gas Protocol 2013). The guidance defined a minimum boundary for types of emissions to include:

- Purchased goods and services from cradle to factory exit (cradle-to-gate)
- All upstream (cradle-to-gate) emissions of purchased capital goods
- Fuel- and energy-related activities not included in scope 1 or scope 2
- Upstream transportation and distribution
- Disposal and treatment of waste generated when not accounted for directly by the organization
- Business travel by employees of the organization in vehicles not owned or operated by the organization
- Employee commuting

---

[7] The GHG Protocol categorizes GHG emissions by "scopes." In their system, direct emissions are scope 1, emissions related to imported electricity, heat, steam, and cooling are scope 2, and other indirect GHG emissions not included in scope 2 are scope 3.

– Upstream leased assets when not reported as scope 1 or scope 2
– Downstream transportation and distribution
– Downstream processing of sold products
– Use of sold products
– End-of-life treatment of sold products
– Downstream leased assets
– Operations of franchisees when not reported by the franchisor
– Investments not included in scope 1 or 2.

A diagram published in the GHG Protocol standards provides a good conceptual presentation of the differences among the three scopes of emissions (Fig. 3.1).

3.5 Quantification at the Product Level

Decarbonizing the world's economy depends upon policy makers, industry, and consumers taking the rising costs of carbon emissions into account and collectively choosing lower-carbon options. Much can be achieved with existing technologies, and yet more will be possible when purchasing decisions are based not only on product performance and price, but also on reduction of a product's carbon footprint. The aviation sector, for example, which is dependent for the foreseeable future on energy-dense liquid fuel to power jet aircraft, has committed to carbon–neutral growth from the year 2020 (ATAG 2021). It plans to achieve this through a series of improvements including weight reductions in aircraft construction, improved engine performance, and the replacement of fossil fuels with alternative aviation fuels. These goals were established long before the "black swan" pandemic event of 2020 sharply curtailed aviation greenhouse gas emissions.

Brazil has led the world in decarbonizing road transport fuel by increasing the use of ethanol produced from sugarcane. In the transport sector, the country has committed to achieve GHG reductions of 37% from a 2005 base by 2025. Industry observers there believe the reductions of GHG emissions could reach 43% by 2030. This will be achieved by increasing the share of sustainable biofuels in the country's transport fuels to approximately 18% by 2030. GHG emission reductions occur because 90% of the emissions associated with the use of petroleum-based gasoline are avoided when substituted by sugarcane-based ethanol (Biofuels International 2015).

3.5.1 Carbon Footprint of Products Based on Life Cycle Assessment

In France, the Grenelle laws of 2009 and 2010 (ADEME 2020) resulted in piloting environmental product labeling programs such as that undertaken by Groupe Casino.

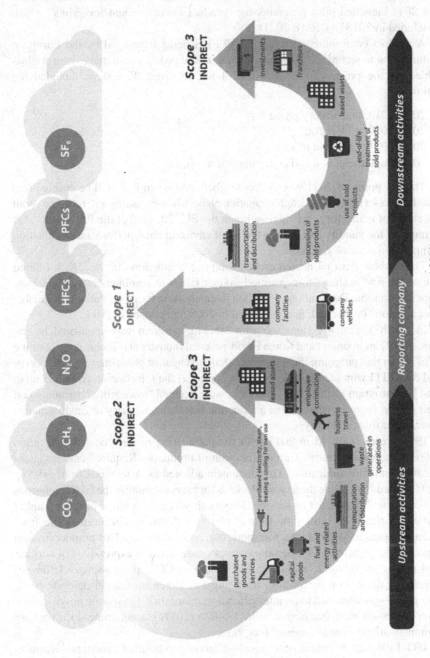

Fig. 3.1 Visual descriptions of GHG protocol scopes 1, 2, and 3 (Greenhouse Gas Protocol 2011)

The clothing, furnishings, and hotel sectors also participated in voluntary initiatives. Following this voluntary initiative in France, the European Commission in 2013 and 2014 launched pilot programs for "product environmental footprints" which concluded in 2018 (ADEME 2021).

A product environmental footprint (PEF) is being developed by the European Commission to record the sustainability impacts of products and make them comparable, but the process in 2021 was still not finalized. Key steps included the following:

- 2008–2013, preparatory phase
- 2013–2019, pilot phase
- 2019–2021, transition phase
- 2021, implementation and communications phase.

The key point for 2021 was to decide where and when PEF will be enacted into law. This was a task for the implementation phase when decisions will be made about the extent of mandatory PEF application in the EU. The goal of the PEF initiative is to improve the validity and comparability of environmental performance evaluation compared to existing methods.

The Carbon Trust pioneered carbon footprint certification in the UK. It offered one of the first business-to-consumer product labels to communicate carbon footprint information. Declarations of carbon intensity were based on Publicly Available Specification (PAS) 2050, first published in 2008 and revised in 2011.

In South Korea, a carbon footprint labeling program was sponsored by the Ministry of Environment and Korea Environmental Industry and Technology Institute (KEITI). In this program, the CFP was calculated against guidelines and then certified by KEITI with annual follow-up to ensure that the information remained up to date. The program identified products that were deemed "low carbon," meaning that their carbon footprint represented a reduction below the Ministry of Environment's baseline for that product type.

ISO entered the field in 2013 with the publication of Technical Specification 14067, Greenhouse gases—Carbon footprint of products—Requirements for quantification and communication. This document adapted LCA methods from existing ISO standards to provide the technical means for users to quantify the GHG emissions associated with a product or service. The result was a per-unit-of-product number that summarized GHG emissions from all phases of the product or service's life, from material inputs to manufacture, transport, use, and end of life. This number is often expressed as grams of CO_2e per kilogram of product. Other expressions of normalized results are possible, including the amount of CO_2e per passenger kilometer traveled or the carbon footprint of attending a conference on a per-delegate basis. ISO 14067 was upgraded to an international standard in 2018, and the provisions on communication were transferred to ISO 14026 (2017a), a guideline document for communication of environmental footprints.

ISO 14067:2018 Greenhouse gases—Carbon footprint of products—Requirements and guidelines for quantification specifies principles, requirements, and guidelines for the quantification and reporting of the carbon footprint of a product (CFP) in a manner consistent with international standards on life cycle assessment. Requirements and guidelines for the quantification of a partial CFP are also specified. This document addresses only a single impact category: climate change. Carbon offsetting and communication of CFP or partial CFP information are outside the scope of this document. This document does not assess any social or economic aspects or impacts, or any other environmental aspects and related impacts potentially arising from the life cycle of a product.

ISO 14067 is based on the LCA principles and requirements contained in ISO 14040 (ISO 2006a) and ISO 14044 (ISO 2006b), standards last revised in 2006. These standards provide a common framework for life cycle assessment practitioners throughout the world. ISO 14067 complements these standards by defining requirements specific to carbon footprints of products (CFPs). The standard differentiates full and partial CFPs. A partial CFP is limited to selected stages or processes within a product's life cycle and is typically developed for business-to-business communication. Different suppliers, for example, may develop partial CFPs for supplied inputs or parts to a customer who will aggregate CFP information from its supply chain to develop a full CFP for a final product. The partial CFP could be based on just the processes employed by the supply chain partner (called a "gate-to-gate" CFP) or may aggregate only the emissions of the stages of the life cycle up to and including the supplier's transformation of inputs ("cradle-to-gate").

ISO 14067 stipulates that, where relevant, CFP product category rules shall be adopted. "Product category rules" are defined as "a set of specific rules, requirements and guidelines for developing Type III environmental declarations and footprint communications for one or more product categories" (ISO 2016). A Type III environmental declaration is one that presents quantified environmental information on the life cycle of a product to enable comparisons between products fulfilling the same function. Declarations of this kind typically are made as "environmental product declarations" meeting the requirements of ISO 14025 (ISO 2006c).

Environmental product declarations differ from CFPs because they take into account a wide range of environmental impact categories rather than just climate change. Some LCA experts believe that taking a holistic view of environmental impacts provides a better basis for decision making than just focusing on a single attribute such as carbon or water. Nonetheless, climate change remains an urgent environmental challenge for humankind in the twenty-first century, with expected impacts from global warming that will change the relationships between humans and their environment in profound ways.

3.6 The Puraglobe Example

In the USA as elsewhere in the world, companies collect and re-refine used lubricating and hydraulic oil. The resulting "base oil" products—the primary component in automotive lubricating oil and industrial hydraulic oils—can meet or exceed quality characteristics for similar conventional products refined from virgin petroleum and reduce carbon emissions by half or more. Customers for these products have included major automotive manufacturers who are motivated to achieve overall reductions in the GHG intensity of their products (Möhr 2019).

Located in an industrial park near the town of Zeitz in the eastern German state of Saxony-Anhalt, an oil refinery named Puraglobe produces base oil and solvents from recycled lubricating and hydraulic oils. This feedstock is collected from industrial and retail sources of used oil and transported by truck and rail to the Elsteraue industrial park. There, the oil is re-refined and separated into multiple grades of lubricating oils and solvents, with the main co-products consisting of naphtha, diesel, and bitumen.

Technology. In 1994, Puraglobe became the world's exclusive licensee for Honeywell UOP's HyLube™ technology process. The Puraglobe refinery in Elsteraue, Germany, opened in 2004 and became the first re-refiner in the world to produce Group III synthetic oils. Successful commercialization of the refinery's output led to a refinery expansion in 2008 and a technology upgrade in 2015 when UOP's newer HyLubeSAT™ process was installed in Puraglobe's second refinery unit. The two units produce Group III synthetic oils according to American Petroleum Institute (API) classifications. A second facility at the Tampa Bay Port in Florida in the USA was planned to increase the production of these synthetic oils. When built, it will become the second refinery in North America (after Petro Canada) to produce Group III lubricants.

Product quality. Puraglobe's lubricants and solvents meet Group II and III API classifications and, using UOP HyLubeSAT™ technology, approach Group IV. Puraglobe has tested its SAE 5W50 lubricant in the demanding conditions of motor car racing. This lubricant helped the "Care For Climate" racing team to secure first place the Nurburgring (Germany) competitions in 2017 and 2018. In association with the British firm Applied Graphene Materials, Puraglobe introduced nanomolecules to its Graphenics® product line for higher performance and better protection against friction and wear.

Carbon footprint. Customers attracted to the idea of using recycled oils prompted Puraglobe to seek the quantification of its product's carbon footprint. It commissioned the Institut für Energie und Umweltforschung (IFEU) in Heidelberg to perform this work. IFEU's study report considered emissions associated with transporting oil from collection points (garages), from the use of raw material and energy inputs, and from the re-refining process. The result was a cradle-to-gate partial carbon footprint that was verified by NSF Certification, LLC. IFEU also calculated—based on a life cycle assessment it previously had carried out on European oil refineries in 2005—the amount of CO_2e savings that Puraglobe achieved through the re-refining used oils compared to the production of base oil from virgin petroleum.

Circular economy contributions. One of the strategies for the transition to a low-carbon economy involves the development of the "circular economy." ISO 20400:2017 Responsible Purchasing—Guidelines defines the term as "an economy that aims to restore and regenerate and tends to preserve the intrinsic value and quality of products, components and materials at every stage of their use, distinguishing biological and technical cycles" (ISO 2017b). The definition comes from the concepts promoted since 2012 by the Ellen MacArthur Foundation[8] to replace the "take, transform, and discard" approach, derived from the production of consumer goods, and replace it with material recovery, recycling, and reuse.

Puraglobe has participated in the circular economy in two ways, one major and one minor. Its major contribution is the recovery of lubrication oil and the regeneration of this recycled material into high-value synthetic oils and other products. Until 2018, the company also applied circular economy concepts by purchasing steam produced by a neighboring facility in its industrial park. A nitric acid production plant there used high-temperature thermal oxidation to destroy dangerous air pollutants. With the heat generated by this operation, the plant produced and sold steam to its neighbor Puraglobe. This relationship ended in 2018 when Puraglobe began sourcing its steam from a newly built combined cycle cogeneration plant installed on its site.

Alternatives to re-refining. According to IFEU's 2005 analysis, the end of life of hydraulic and lubricating oils in Germany consisted of the following: 40% of the volume of oil is lost during the use phase, 7% is used directly as fuel, 18% is regenerated and burned as fuel, and 35% is regenerated into synthetic oil. More recent figures are not available, although the existence of Puraglobe on the market and its success in re-refining used oils since 2004 should have increased the proportion of recycled waste oils and decreased the proportion of thermally destroyed waste oils.

Carbon footprint study conclusions. Puraglobe in Germany demonstrated the feasibility of recovering and re-refining used oils. It has plans to expand this business model in the USA with the construction of a new plant using the latest HyLubeSAT™ technology in Tampa, Florida. Like the one operating in Germany, the North American plant should recover used oils and re-refine them to make products that are of far greater value than that of used oil consumed as fuel. Customers of lubrication and hydraulic oils therefore have the option to purchase a product with a carbon footprint lower than that of virgin petroleum oils that meets high-quality synthetic oil standards.

3.7 Who Should Participate in Counting Carbon?

Carbon accounting in an organization involves mobilizing a variety of functions:

– Operational functions
– Crosscutting functions
– Management functions.

For direct emissions and indirect emissions from imported energy, personnel at an organization's headquarters may consolidate information relating to direct and indirect emissions collected and aggregated from the following types of facility-level personnel:

---

[8] Foundation headed by Dame Ellen MacArthur, renowned British winner of sailing races, active in the promotion of environmental protection.

1. Production planning staff for estimates of energy use and process emissions
2. Operational managers for records of actual energy consumed associated manufacturing process emissions
3. Purchasing department staff for imported electricity, heat, and cooling invoices
4. Maintenance personnel for records of installed measurement devices including their maintenance and calibration
5. Engineering department personnel for analyses of process changes that can reduce greenhouse gas emissions.

Organizations should model the efficiency of energy use per unit of product and benchmark it to best-in-class technologies. They should also strive for the best achievable measurement accuracy—generally targeting no more than $\pm 2\%$ variance from measured values associated with direct emissions. Emissions that are modeled may present higher degrees of uncertainty but should remain within an uncertainty range of not more than 5%.

For indirect emissions, a wider set of organizational resources will need to be mobilized:

1. Real estate and property management personnel
2. Staff responsible for the company's travel planning and purchasing
3. Personnel who receive visitors to the organization
4. Personnel who contract for external service provision
5. Accounting personnel
6. Staff who procure logistics services (transport, delivery, warehousing, etc.).

3.8   Environmental Management System

ISO 14001:2015 provides an ideal framework for implementing carbon accounting. The 2015 revision of the standard introduced a requirement to consider a life cycle perspective when identifying the organization's environmental aspects and when planning and controlling its operations. The boundaries for reporting emissions from indirect sources show how much carbon accounting is informed by and takes into account life cycle considerations.

An effective environmental management system is based on thoughtful systems of control which must embrace carbon accounting. The design and implementation of controls should consider:

1. Policies to reduce the carbon footprint of activities which are supported by management at the highest level (responsibilities)
2. The review of GHG data when analyzing environmental aspects during both initial analysis and subsequent review, and the establishment of thresholds of significance (categorization and prioritization)
3. Organization-wide planning to reduce GHG emissions wherever possible
4. Actions that can be taken to reduce or mitigate GHG emissions (implementation)
5. Checking related to actions taken and results achieved (internal audits)

6.   Tracking and reporting of actions that reduce the carbon footprint (management review and continual improvement).

3.8.1 Greenhouse Gas Reporting According to ISO 14064 Part 1

The normal outcome of GHG inventory management is a statement of the organization's GHG emissions and removals, also known as a GHG inventory report. The full assessment of an organization's GHG emissions and removals may be developed for internal use only, or it may be communicated to individuals considered by the organization to be its "intended users."

> **ISO 14064:2018 Part 1[9]**
> **3.4.4 intended user**
>
> "Individual or organization (3.4.2) is identified by those reporting GHG-related information as being the one who relies on that information to make decisions.
>   "Note 1 to entry: The intended user can be the client (3.4.5), the responsible party (3.4.3), the organization itself, GHG program (3.2.8) administrators, the financial community, or other affected interested parties, such as local communities, government departments, general public, or non-governmental organizations."

ISO 14064 Part 1 does not specify how organizations should report their GHGs, but it does suggest a format. In addition, organizations that report their GHG emissions and removals are not required to make them public. However, if they claim conformity to ISO 14064 Part 1, or if they have their reports verified, they must prepare a report (ISO 2018b).

Many organizations now make their GHG inventories public by reporting to CDP or to other platforms. In most cases, organizations that publicly report their GHG emissions and removals obtain the services of an independent third-party audit organization to ensure that quantified emissions are "fairly stated" and reported in conformity with criteria such as ISO 14064 Part 1 or the GHG Protocol and any applicable GHG program rules.

There is a growing trend for increased transparency with respect to corporate emissions reporting. Companies should plan emissions inventory reports to meet the information needs of their interested parties. In most cases, this means responding to GHG information requests. These may include:

---

[9] Definition ©ISO. This material is reproduced from ISO 14064-1:2018, with permission of the American National Standards Institute (ANSI) on behalf of the International Organization for Standardization. All rights reserved.

– Reporting to the organization's management about total emissions and removals or product carbon intensity for strategic planning and environmental program development
– Fulfillment of legal requirements through reporting to regulatory authorities
– Informing legal staff responsible for drafting mandatory disclosures to shareholders
– Providing information to public relations and marketing staff to communicate on environmental, corporate social responsibility (CSR), or Sustainable Development Goals and results
– Making voluntary disclosures to GHG programs.

In planning its GHG inventory report, the organization should adapt the content of its GHG statements to the needs of its intended users. The scope and content of the report may be influenced by the organization's GHG policies, strategies, and programs. This will influence the frequency and format of the GHG report (ISO 2018b).

When an organization chooses to issue a publicly available statement of its GHG emissions and removals and wishes to claim compliance with ISO 14064 Part 1, the organization is required to include the following elements in its GHG inventory report (ISO 2018b, 9.3.1)[10]:

– A description of the organization
– Identification of the person(s) responsible for writing the report
– The period covered by the report
– Documentation of the organizational and reporting boundaries (5.1)
– Discussion of the criteria applied by the organization to determine which emissions are significant
– A list of direct GHG emissions, quantified separately for CO_2, CH_4, N_2O, NF_3, SF_6, and other appropriate GHG groups (HFCs, PFCs, etc.), in tons of CO_2e (5.2.2)
– A description of how biogenic CO_2 emissions and removals are treated in the GHG inventory and how relevant biogenic CO_2 emissions and removals are quantified separately in metric tons of CO_2e (see Appendix D)
– If quantified, direct GHG removals in metric tons of CO_2e (5.2.2)
– Justification for excluding quantification of any significant GHG source or sink (5.2.3)
– Indirect GHG emissions, reported by emission category, in metric tons of CO_2e (5.2.4)
– The chosen historical base year and the GHG inventory for the base year (6.4.1)
– Explanation of any changes in the base year or other historical GHG data, and any recalculation of the base year or other historical GHG inventory (6.4.1), and documentation of any comparability limitations resulting from this recalculation

---

[10] Clause 9.3.1 ©ISO. This material is reproduced from ISO 14064-1:2018, with permission of the American National Standards Institute (ANSI) on behalf of the International Organization for Standardization. All rights reserved.

- A reference to quantification approaches or their description, specifying the reasons for their selection (6.2)
- Explanation of any changes to the quantification approaches previously used (6.2)
- Reference to, or documentation of, GHG emission or removal factors used (6.2)
- Description of the impact of uncertainties on the accuracy of GHG emissions and removal data by category (8.3)
- Description and results of the uncertainty assessment (8.3)
- A statement certifying that the GHG report was developed in accordance with Part 1
- A disclosure indicating whether the inventory, report, or GHG declaration has been verified, including the type of verification and the level of assurance achieved
- GWP values used in the calculations, as well as their source; if the GWP values were not taken from the latest IPCC report, an indication of the emission factors or database reference used in the calculations, as well as their source.

ISO 14064 Part 1 also recommends including the following optional elements in a GHG inventory report (ISO 2018b, 9.3.2)[11]:

- A description of greenhouse gas policies, strategies, and programs
- If appropriate, a description of GHG reduction initiatives that help reduce emissions or increase removals, including those that occur outside the organization's operational or reporting boundaries, quantified in metric tons of CO_2e (7.1)
- If appropriate, purchased or developed GHG emission reductions and removal enhancements from GHG emission reduction and removal enhancement projects, quantified in tons of CO_2e (7.2)
- As appropriate, a description of applicable GHG program requirements
- GHG emissions or removals disaggregated by facility
- Total quantified indirect GHG emissions
- A description and presentation of additional indicators, such as efficiency or GHG intensity of emissions (emissions per unit of production) ratios
- Assessment of performance against appropriate internal and/or external benchmarks
- Description of GHG management and monitoring procedures
- GHG emissions and removals from the previous reporting period
- If appropriate, an explanation of GHG emissions differences between the present inventory and the previous one.

All these elements are developed in clauses 9.3.1 or 9.3.2 of the standard, mainly for the internal uses of organizations. Many organizations or GHG programs to which organizations report follow a more streamlined format in their internal and external reporting.

---

[11] Clause 9.3.2 ©ISO. This material is reproduced from ISO 14064-1:2018, with permission of the American National Standards Institute (ANSI) on behalf of the International Organization for Standardization. All rights reserved.

ISO 14064 Part 1 does suggest a format in Annex F for the organization of quantified information to be included in a report. The main elements in this format include[12]:

1. A general description of the organization's objectives and inventory objectives
2. Organizational boundaries
3. Reporting boundaries
4. Inventory of quantified GHG emissions, removals, and storage
5. An accounting of carbon financial instruments and GHG emissions reduction initiatives.

The choice of inventory information (boundaries and level of detail) and reporting accuracy are the responsibility of management. Disclosure requirements for publicly traded companies may differ from those of private companies. We develop in further detail the theme of public reporting to CDP in Chap. 4.

Questions for Readers

1. Why is it important for an organization to set a boundary (organizational, reporting) for its emissions inventory?
2. What is the difference between GHG emissions and GHG removals?
3. Why is it important to involve a variety of organizational functions and personnel in the quantification of greenhouse gas emissions?
4. What are the characteristics of an environmental management system that make its implementation useful to organizations that quantify their greenhouse gas emissions?
5. What is the value to consumers of requiring disclosures of product environmental footprint information on products sold at retail?

References

ADEME (2020) Environmental labeling: context and objectives. https://www.ademe.fr/expertises/consommer-autrement/passer-a-laction/reconnaitre-produit-plus-respectueux-lenvironnement/dossier/laffichage-environnemental/affichage-environnemental-contexte-objectifs. Accessed 2 Feb 2020
ADEME (2021) Environmental labeling: work history and feedback. https://www.ademe.fr/expertises/consommer-autrement/passer-a-laction/reconnaitre-produit-plus-respectueux-lenvironnement/dossier/laffichage-environnemental/affichage-environnemental-historique-travaux-retours-dexperience. Accessed 20 July 2018

---

[12] Annex F ©ISO. This material is reproduced from ISO 14064-1:2018, with permission of the American National Standards Institute (ANSI) on behalf of the International Organization for Standardization. All rights reserved.

ATAG (2021) Climate change. https://www.atag.org/our-activities/climate-change.html. Accessed 14 Feb 2014

Biofuels International (2015) The role of cleaner fuels after COP21. https://biofuels-news.com/article_display/?volume=10&issue=1&content_item=1005. Accessed 14 Feb 2016

Climate Registry (2020) Climate registry default emission factors. https://www.theclimateregistry.org/wp-content/uploads/2020/04/The-Climate-Registry-2020-Default-Emission-Factor-Document.pdf. Accessed 15 Jan 2021

Greenhouse Gas Protocol (2004) Corporate accounting and reporting standard, revised edition. https://ghgprotocol.org/corporate-standard. Accessed 7 June 2021

Greenhouse Gas Protocol (2011) Corporate value chain (Scope 3) Accounting and reporting standard https://ghgprotocol.org/sites/default/files/standards/Corporate-Value-Chain-Accounting-Reporing-Standard_041613_2.pdf. Accessed 30 May 2021

Greenhouse Gas Protocol (2013) Technical guidance for calculating scope 3 emissions. https://ghgprotocol.org/sites/default/files/standards/Scope3_Calculation_Guidance_0.pdf. Accessed 17 Jan 2021

Greenhouse Gas Protocol (2015) Scope 2 guidance. https://ghgprotocol.org/scope_2_guidance. Accessed 17 Jan 2021

IPCC (2021) Task force on national greenhouse gas inventories—FAQs. https://www.ipcc-nggip.iges.or.jp/faq/faq.html. Accessed 15 Dec 2020

ISO (2006a) ISO 14040:2006 environmental management—life cycle assessment—principles and framework. https://www.iso.org/standard/37456.html. Accessed 8 Feb 2021

ISO (2006b) ISO 14044:2006 environmental management—life cycle assessment—requirements and guidelines. https://www.iso.org/standard/38498.html. Accessed 8 Feb 2021

ISO (2006c) ISO 14025:2006 environmental labels and declarations—type III environmental declarations—principles and procedures. https://www.iso.org/standard/38131.html. Accessed 8 Feb 2021

ISO (2015) ISO 14001:2015 environmental management systems—requirements with guidance for use. https://www.iso.org/standard/60857.html. Accessed 7 June 2021

ISO (2016) ISO 14067:2018 Greenhouse gases—carbon footprint of products—requirements and guidelines for quantification. https://www.iso.org/standard/71206.html. Accessed 8 Feb 2021

ISO (2017a) ISO 14026 environmental labels and declarations—principles, requirements and guidelines for the communication of footprint information. https://www.iso.org/standard/67401.html. Accessed 10 June 2021

ISO (2017b) ISO 20400:2017 responsible purchasing—guidelines. https://www.iso.org/standard/63026.html. Accessed 14 May 2021

ISO (2018a) ISO 50001:2018, energy management systems—requirements with guidance for use. https://www.iso.org/standard/69426.html. Accessed 7 June 2021

ISO (2018b) ISO 14064-1:2018 greenhouse gases—part 1: specification with guidance at the organization level for quantification and reporting of greenhouse gas emissions and removals. https://www.iso.org/obp/ui/#iso:std:iso:14064:-1:ed-2:v1:en. Accessed 18 Jan 2021

ISO (2018c) ISO 14067:2018 greenhouse gases—carbon footprint of products—requirements and guidelines for quantification. https://www.iso.org/standard/71206.html. Accessed 15 Jan 2021

ISO (2019) ISO 14033 environmental management—quantitative environmental information—guidelines and examples. https://www.iso.org/standard/71237.html. Accessed 7 June 2021

Möhr S (2019) Interview with Sönke Möhr, Director of global sales, marketing and communication, Puraglobe Germany GmbH

Chapter 4
Reducing Emissions

4.1 Carbon Markets and GHG Mitigation

The flexible mechanisms established after Kyoto resulted in the establishment of the Clean Development Mechanism and similar programs designed to reduce greenhouse gas (GHG) emissions and increase GHG removals. Such initiatives "mitigate" global warming by reducing emissions of CO_2 and other GHGs into the atmosphere or by increasing carbon dioxide (CO_2) sequestration in such sinks as forests and soil. But despite the efforts, total global annual GHG emissions, including from land-use change, reached a record high of 59.1 $GtCO_2$-e in 2019, up 55% since 1990 (UN Environment Programme 2020).

Despite awareness of the risks posed by climate change, and the negotiation of the Kyoto Protocol in 1997, the quantities of GHGs emitted to the atmosphere from 1990 to 2020 represent about half of all GHGs emitted since the beginning of the industrial revolution (Stainforth 2020). The data show that efforts in the first two decades of the twenty-first century to mitigate GHG emissions have been insufficient. Instead, the climate crisis facing the world has only grown worse with worldwide GHG emissions continuing to rise (Millington et al. 2020, p 28). This bleak picture would be even more grim without projects to reduce emissions and enhance removals which is the subject of this chapter (Fig. 4.1).

There is considerable variety in the types of mitigation projects. This is a strength, because it is important that all economic sectors contribute to decarbonizing the world's economies. Projects that result in the issuance of tradable carbon credits have the following characteristics in common:

- The project is developed in accordance with rules specified by a "greenhouse gas program" such as the CDM or many other voluntary or regulatory programs that have since been established around the world
- The rules of the program specify requirements for:
  - Recognition of the project
  - Monitoring of GHG sources, sinks, and reservoirs

J. C. Shideler and J. Hetzel, *Introduction to Climate Change Management*, Springer Climate, https://doi.org/10.1007/978-3-030-87918-1_4

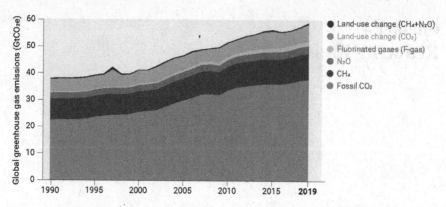

Fig. 4.1 Global GHG emissions from all sources (UN Environment Programme 2020)

- – Quantification of emission reductions and removal enhancements
- – Validation and verification of emission reductions and removal enhancements
- – Issuance of carbon credits

- Carbon credits are assigned serial numbers and held in a registry
- Procedures exist to prevent the double claiming of emission reductions and removal enhancements.

Administering a carbon credit registry is complex. Many registries serve specific regulated markets in particular countries, others cater to a wider universe of carbon credit buyers and sellers. In Europe, a common market existed for European Union Emissions Trading System (EU ETS) allowances and, until 2020, for CDM credits. In North America, the regulated carbon markets of the subnational jurisdictions of California and Quebec are linked. Most registries have operated without significant problems since their founding in 2005 or later. In a notable exception, a value-added tax (VAT) fraud in the EU rocked the carbon markets in 2009. "Between 2008 and 2009, €1.6 billion were swindled in a huge carbon quota market scam dubbed the 'fraud of the century'." The fraud was discovered and prosecuted (Boyer Kind 2018). The fraud was based on a simple mechanism of buying and selling false carbon allowances in the market, the cost of which was paid for through VAT refunds, a scam made possible by the lack of European fiscal harmonization.

4.1.1   The Concept of Additionality

Because the point of an emission reduction or removal enhancement project is to create a financial instrument called a "carbon credit" that represents a ton of CO_2e reduced or removed, it is important that the created emission reduction or removal enhancement would not have occurred anyway, without the existence of the project. The term "additionality" serves to distinguish projects that deserve carbon crediting from those that can be considered "business as usual." In other words, a project that is "additional" is one that would not have been developed absent the incentives provided by the earning of carbon credits. Additionality is determined through an analysis of baseline scenarios.

An International Standard, ISO 14064 Part 2, Greenhouse gases—Specification with guidance for monitoring, quantifying, and reporting greenhouse gas emission reductions and removal enhancements at the project level, defines the basic concepts used in GHG project accounting. Its rules typically are supplemented by rules adopted by various voluntary and regulatory GHG programs. A key term in determining additionality is "baseline scenario," which the standard defines as follows:

ISO 14064:2019 Part 2[1]
3.2.6 baseline scenario

"Hypothetical reference case that best represents the conditions most likely to occur in the absence of a proposed GHG project."

Project developers are required to identify at least two plausible scenarios:

- The activity proposed for the project continues with no change from past practice ("business as usual")
- The activity is modified by implementation of the project, which results in GHG emission reductions or removal enhancements.

Concretely, this could mean in the case of a landfill that methane (CH_4) continued to be released to the atmosphere as a fugitive emission in the absence of a project to collect and combust it. Or a forest continued to be managed without the implementation of sustainable management practices and reduction in harvesting. If the project scenario is not common practice, or if economic or technological barriers make its implementation infeasible or economically unattractive, then implementing the project is determined to be "additional."

The selection of a plausible "baseline scenario" enables the project developer to calculate the project's "baseline." The project baseline is defined in ISO 14064 Part 2 as "quantitative reference(s) of GHG emissions (3.1.5) and/or GHG removals (3.1.6) that would have occurred in the absence of a GHG project and provides the basis for comparison with project GHG emissions and/or GHG removals" (ISO 2019a).

The following two graphics illustrate how emission reductions and removal enhancements may be quantified with respect to a baseline (Figs. 4.2 and 4.3):

The analysis of baseline scenarios and resulting project baselines can be quite complex. In a Reduced Emissions from Deforestation and Degradation (REDD) project, the baseline assumes the continued loss of forest carbon due to continued

---

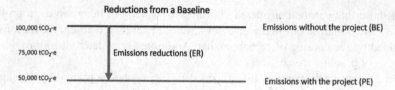

Fig. 4.2 GHG reductions from a baseline

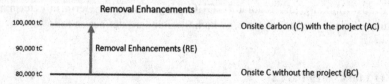

Fig. 4.3 CO$_2$ removal enhancements

deforestation and degradation. Project activities are designed to reduce the loss of those carbon stocks. This means that the maintenance, or enhancement, of forest carbon stocks above the baseline are additional to the business-as-usual scenario. This hypothesis is illustrated in Fig. 4.4.

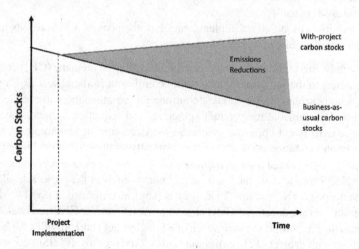

Fig. 4.4 Process to implement forest management under REDD program (Tigray Regional State 2015)

4.1.2 Obtaining Recognition

There are two main ways that an emission reduction or removal enhancement project obtains "recognition." As described above, the CDM Executive Board approved universally applicable methodologies to reduce GHG emissions or enhance removals. It required that each individual instance of a project be "validated" by a third party to ensure that resulting emission reductions or removal enhancements would be "additional" and not result from "business as usual." This approach has been adopted by some voluntary GHG programs such as the Verified Carbon Standard (VCS) and the American Carbon Registry (ACR).

Another approach to ensure the additionality of projects is to establish "performance-based" protocols. Under this approach, a GHG program surveys common practice and technological and economic barriers in a particular geography and determines the conditions under which a project may be deemed additional. These conditions are included as eligibility rules in a project protocol. If a project meets them, its emission reductions or removal enhancements receive carbon credits after successful verification by a third party. This approach is used in North America by the California Air Resources Board, by the Quebec Ministry of Sustainable Development and Fight Against Climate Change, and, in the voluntary market, by the Climate Action Reserve.

Both approaches lead to the "recognition" by the relevant GHG program of the activities undertaken by a project developer to reduce emissions or enhance removals of greenhouse gases.

4.1.3 Monitoring of GHG Sources, Sinks and Reservoirs

An important task for project developers is to select, install, and operate monitoring procedures and technology that will provide sufficient evidence that emission reductions and removal enhancements have occurred during a reporting period. The evidence must be robust enough to persuade a third-party verifier that the rules of the applicable methodology or protocol have been met and that the claimed number of carbon credits have been legitimately earned. The types of technologies deployed depend upon the circumstances of each project.

Many projects involve the flow of gases, such as landfill gas, biogas in a dairy digester, or the "tail gas" of a nitric acid plant. For these types of projects, the main monitoring technology consists of a flow meter to quantify the volume (or mass) of a gas, and a gas analyzer to profile how much of a specific gas is present in the gas stream. These meters are connected to data recorders and computers that use communications protocols to transmit data to a control room or to a remote office. Recorded data are compiled and checked, and equations found in the approved methodology or protocol quantify GHG emission reductions and removal enhancements.

Forestry projects operate quite differently. Project methodologies and protocols have largely adopted the same approach used by forest owners to calculate how much merchantable timber their lands contain. The values obtained provide the data for quantifying how much CO_2 their trees have sequestered. This process involves taking an inventory of forest timber by establishing sampling plots at randomly selected locations and then measuring all the trees within that sampling plot. Each tree species is identified, and its diameter at breast-height is measured with a logger's tape. The height of the tree is calculated with a laser clinometer, and any defects or cavities in the tree are recorded. From these data, the forester uses allometric equations to calculate the amount of carbon contained in each tree. The amount of carbon in trees in the sampling plot is summed, and the results of all the sampling plot carbon stock calculations are extrapolated to the whole forest. Light detection and ranging—LiDAR—is an alternate technology that can replace on-ground inventory stock-taking but is not universally accepted by carbon offset registries.

4.1.4   Quantification of Emission Reductions and Removal Enhancements

The most common equation used in mitigation projects is $ER = BE - PE$, where "ER" stands for emission reductions, "BE" for baseline emissions, and "PE" for project emissions. As discussed above, a baseline in a mitigation project represents the estimated emissions that would have occurred absent the implementation of the project. So, it is logical to conclude that the emission reductions achieved will equal the delta between the baseline and actual monitored emissions that occurred during the project's reporting periods. Project emissions in a landfill gas project typically include CO_2 from the combustion of fuels associated with project activity and other incidental emissions associated with the project such as electricity consumed to run the landfill's collection system blower and monitoring equipment. The main source of emission reductions comes from the destruction of methane which has a global warming potential (GWP) of 25 compared to the GWP of 1 for CO_2. In other words, total emission reductions from a landfill gas project will nearly amount in each reporting period to 25 times the amount of CO_2 emitted by a flare or other combustion device.[2]

The quantification of removal enhancements in a forestry project is more complex and uses a different equation. As indicated above in our discussion of baselines, the final equation for removal enhancements could be $RE = (AC - BC) \times 44/12$, where "RE" stands for removal enhancements, "AC" for actual carbon at the end of the reporting period, and "BC" for baseline carbon. The term 44/12 is included in the equation to convert molecular carbon to carbon dioxide gas. Molecular carbon is produced by chemical reactions during photosynthesis where carbon dioxide absorbed by the tree is stored as carbon in the wood of the tree.

---

[2] For this example, we assume the use of GWPs from the IPCC's Assessment Report 4.

The forester takes many steps before applying this final equation. The purpose of measuring trees in a sampling plot is twofold. First, the forester wishes to know the total amount of stored carbon in the forest's trees. This information is also used by the forest owner to evaluate how much timber could be produced if harvesting were to occur, or how much the forest is worth if it were to be sold. For removal enhancement projects, the key variable to calculate is the amount of change in the carbon stocks from one reporting period to another. This is the flux in stored carbon that receives carbon credits, rather than the total stock of stored carbon.

To calculate carbon stocks, forests are stratified by vegetation type and geography—variables that influence rates of growth. The forester uses a "growth-and-yield" model to calculate the expected change in stored carbon at points in time. For example, a physical inventory may have been taken of trees in sampling plots a year or more before the end of a reporting period. Using a growth-and-yield model, the forester can "grow" the forest to the end of the reporting period and report on the results. In the case of removal enhancement projects, project protocols require the taking of a physical inventory at specified intervals to "true up" the estimated fluxes in carbon storage with those calculated through measurements.

4.1.5 Validation and Verification

Central to the issuance of carbon credits is validation and verification, conducted by competent third-party validators and verifiers.

Validation, according to ISO 14064 Part 3, Greenhouse gases—Specification with guidance for the verification and validation of greenhouse gas statements, is a "process to evaluate the reasonableness of the assumptions, limitations, and methods that support a statement about the outcome of future activities" (ISO 2019b). It occurs either before a mitigation project is implemented, or shortly afterward. The purpose of validation is to provide an opinion, based on a project design document, that the project as planned is likely to produce the intended emission reductions or removal enhancements, and is likely to meet the applicability requirements contained in the selected methodology.

Validation is performed by a validation team using the process requirements described in ISO 14064 Part 3 clause 7. The team's report forms the basis for conclusions and an opinion that is submitted to the GHG program for acceptance. A successful validation provides a project developer with a "green light" to proceed with project implementation and the generation of emission reductions or removal enhancements. As described above, the validation step is not required for projects applying a performance-based protocol such as those approved by the Climate Action Reserve or the cap-and-trade regulation of the California Air Resources Board.

Verification is backwards looking and occurs at selected intervals after project validation. The GHG program or the project developer defines reporting periods when the project has monitored and quantified emission reductions or removal enhancements. The information being verified is therefore historical in nature. Verifiers determine that emission reductions and removal enhancements actually occurred on the basis of evidence gathered and reviewed. Verifiers also check to make sure that the information provided by the person responsible for quantifying the emissions reductions or removal enhancements (the "responsible party") was complete and accurate, properly classified, and correctly assigned to the stated reporting period. Responsible party statements on emission reductions and removal enhancements are typically extensive and complex, and it is rare that verifiers do not discover some "misstatements." Most often these misstatements are relatively minor and below what is called the "threshold of materiality."[3]

ISO 14064-3: 2019 Key Definitions[4]

3.6.15 misstatement "errors, omissions, misreporting, or misrepresentations in the GHG statement (3.4.3)."

3.6.17 material misstatement "individual misstatement (3.6.15) or the aggregate of actual misstatements in the GHG statement (3.4.3) that could affect the decision of the intended users (3.2.4)."

3.6.9 materiality "concept that individual misstatements (3.6.15) or the aggregation of misstatements could influence the intended users' (3.2.4) decisions."

The output of the verifier's work is a verification *opinion*, which ISO 14064 Part 3 (2019b) defines as a "formal written declaration to the intended user (3.2.4) that provides confidence on the GHG statement (3.4.3) in the responsible party's (3.2.3) GHG report (3.4.2) and confirms conformity with the criteria (3.6.10)." Despite the work performed by the verifier, the opinion stops short of providing "absolute" confidence to the intended user of the responsible party's statement that the statement is complete and accurate. Instead, the verifier opines that the statement is a "fair representation" of the achieved emission reductions or removal enhancements. This phrasing is appropriate because the verifier will not examine every possible piece of evidence that supports the responsible party's statement. Intended users of the responsible party's statements understand that some misstatements may remain undetected or uncorrected. This is acceptable as long as they do not exceed, on an aggregated basis, the agreed threshold of materiality.

The approach to verification described in ISO 14064 Part 3 was inspired from financial accounting standards for assurance engagements of historical non-financial information. In this respect, financial auditors and GHG verifiers "speak the same

---

[3] Definitions used in ISO documents are publicly available on ISO's Online Browsing Platform (ISO 2020).

[4] Definitions ©ISO. This material is reproduced from ISO 14064-3:2019, with permission of the American National Standards Institute (ANSI) on behalf of the International Organization for Standardization. All rights reserved.

language," except that the one's focus is on financial accounts while the other's is on statements of carbon dioxide–equivalent tons.

4.1.6 Issuance of Carbon Credits

The "greenhouse gas program" under whose rules and procedures the project was developed and implemented issues carbon credits upon receipt of verification statements for specific reporting period(s),

> **Key Definition**[5]
> **3.2.1 GHG program** "voluntary or mandatory international, national, or subnational system or scheme that registers, accounts, or manages GHG emissions (3.3.2), GHG removals (3.3.4), GHG emission reductions (3.4.8), or GHG removal enhancements (3.4.9) outside the organization (3.2.2) or GHG project (3.4.1).
>
> Note 1 to entry: In this document, a GHG program may also register, account, or manage GHG emissions, GHG removals, GHG emission reductions, or GHG removal enhancements from products (ISO 2019b)."

The 1997 Kyoto Protocol, which was negotiated under the United Nations Framework Convention on Climate Change (UNFCCC), established the first carbon trading mechanisms to address climate change. The Kyoto Protocol led to the establishment in Europe of the EU ETS which allowed regulated entities to fulfill their compliance obligations by retiring Certified Emission Reductions (CERs) issued by the Clean Development Mechanism executive board and Emission Reduction Units (ERUs) issued by countries such as Russia who participated with other industrialized countries in emission reduction programs under the Joint Implementation (JI) mechanism. Support for trading mechanisms to mitigate pollution came from the successful implementation in the USA of an acid rain program that relied on the trading of market allowances to reduce sulfur dioxide and oxides of nitrogen emissions from power plants. The program was adopted by the US Congress in its Clean Air Act Amendments legislation in 1995.

The first regulatory cap-and-trade program in the USA was established by seven northeastern and mid-Atlantic states in 2006. The Regional Greenhouse Gas Initiative (RGGI) program capped emissions at electricity power generating stations. Each participating state adopted RGGI's "model rule" and issued regulations to bring the system into force in their state. California's Air Resources Board followed in 2009 with a program affecting a much wider swath of industrial activity.

---

[5] See Footnote 4.

The California Climate Action Registry had established a voluntary market for carbon credits in the first decade of the 2000s. Later renamed the "Climate Action Reserve," it developed performance-based protocols and issued voluntary carbon credits known as "Climate Reserve Tons" (CRTs, pronounced "carrots"). The Voluntary Carbon Standard—later renamed the Verified Carbon Standard—was launched in 2007 to apply CDM-approved methodologies to voluntary emission reduction and removal enhancement projects anywhere in the world. (The CDM itself only recognized projects implemented in developing countries.) British Columbia, Alberta, and Quebec also developed programs and began to issue carbon credits. In 2013, California and Quebec linked their markets to improve liquidity for trading (Peteritas 2013).

4.2 Project Development: An Overview

A key person in the development of a mitigation project is the "project proponent," also called the "project developer" in some GHG programs. According to ISO 14064 Part 2, the project proponent is the "individual or organization that has overall control and responsibility for a GHG project" (ISO 2019a). Assuming the project is implemented for the purpose of obtaining carbon credits, the project proponent is responsible for obtaining recognition from a GHG program that operates a registry and issues credits. The project proponent implements the project, obtains validation, and verification services, and submits the required documentation to the GHG program. As explained above, a validation phase may not apply if the project is developed under a performance-based protocol.

The following figure[6] provides a typical timeline for the phases of project development and crediting (Fig. 4.5):

Note that some phases in this timeline may not be required by the program criteria under which a project is developed, and some phases are not described in ISO 14064 Part 2. For example, CDM projects require that interested parties be consulted before and after establishing the GHG project plan. This is not a requirement of Part 2. Overall, however, the figure accurately describes the required steps involved in planning and implementing a GHG project.

4.3 Mitigation of Greenhouse Gases

GHG mitigation comes in two forms: reductions of emissions achieved at the organizational level, and emission reductions and removal enhancements achieved at the

---

[6] Figure ©ISO. This material is reproduced from ISO 14064-2:2019, with permission of the American National Standards Institute (ANSI) on behalf of the International Organization for Standardization. All rights reserved.

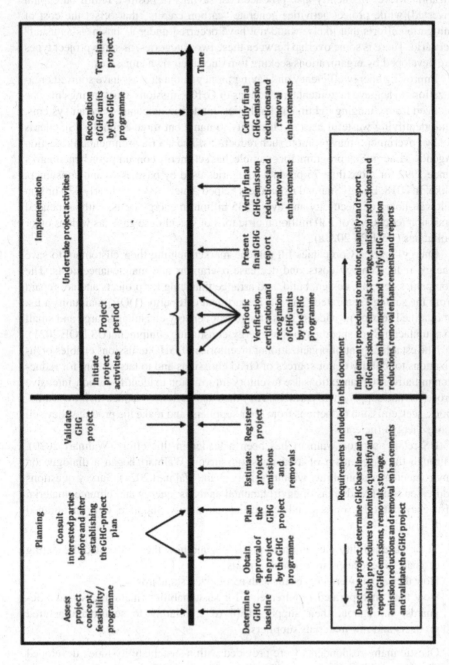

Fig. 4.5 Steps included in a GHG project according to ISO 14064 Part 2 (ISO 2019a, Fig. 4.2)

project level. The main difference is that voluntary mitigation efforts at the organizational level frequently also produce cost savings (a positive return on capital invested) while project activities generate "carbon credits" that offset the cost of mitigation efforts that likely would not have occurred under a "business-as-usual" scenario. There is some overlap between these two categories, as some project types are developed by organizations seeking to reduce their own emissions.

Improving energy efficiency and saving money in the process have spurred organizations to achieve incremental reductions of GHG emissions. Cost savings may be realized from changing lighting technologies, modernizing compressed air systems, and identifying wasteful uses of electricity to name but three examples. Standards set by governments may promote such reductions. The US Environmental Protection Agency's EnergyStar program, for example, has set energy consumption benchmarks since 1992 for more than 75 product categories used by businesses and consumers alike. In 2018, EnergyStar and its partners helped Americans save nearly 430 billion kilowatt-hours of electricity and avoid \$35 billion in energy costs, with associated emission reductions of 330 million metric tons of greenhouse gases as well as other pollutants (US EPA 2021a).

Energy Service Companies fill a niche market helping their customers to save energy, reduce energy costs, and decrease operations and maintenance costs. The companies develop, design, build, and arrange financing for projects and are repaid from the savings generated. The US Department of Energy (DOE) maintains a list of qualified service providers. These range from energy consultants large and small to manufacturers of specific types of energy-consuming equipment (US DOE 2021).

The establishment of organizational inventories of GHG emissions enables organizations to identify major sources of GHG emissions and to target them for reduction initiatives. Organizations use inventory information to identify carbon-intensive processes and supply chain partners. They then take actions to reduce their own emissions, seek emission reductions from their suppliers, and make the products they sell more energy efficient.

US retail operator Walmart has been a leader in this effort (Walmart 2020). Through the mechanism of a 15-question survey, Walmart began a dialogue on environmental sustainability with its supply chain(Walmart 2021). Survey questions addressed four broad areas of environmental analysis: energy and climate, material efficiency, natural resources, and people and community. Suppliers were encouraged to describe:

- Their sustainability measurement processes and how these informed the setting of goals to reduce environmental impacts
- How their goals were consistent with recognized standards
- How they incorporated requirements for sustainability (including social safeguards) throughout their supply chain (e.g., relating to semi-manufactured products and raw materials suppliers)

Questionnaire respondents were provided with a sustainability index developed by The Sustainability Consortium (TSC), a global non-profit organization, and asked to undertake the following activities:

(1)   Establish energy, carbon, water and waste baselines (environmental impact baselines)
(2)   Establish impact reduction goals, align environmental impact data with broader business operations, and set reduction goals to drive both cost and environmental outcomes
(3)   Choose and implement initiatives that will yield the greatest progress against the goals
(4)   Measure progress of reduction initiatives relative to the baseline and goals using the Greenhouse Gas Protocol or Global Reporting Initiative methodologies
(5)   Convey the progress to external stakeholders through company sustainability reports and external venues such as CDP or Ceres

TSC collected supplier data to incentivize Walmart suppliers to deliver more energy savings, environmental or sustainable consumer products (TSC 2019). TSC members and partners include manufacturers, retailers, suppliers, service providers, NGOs, civil society organizations, governmental agencies and academics. TSC convenes diverse stakeholders to work collaboratively to build science-based decision tools and solutions that address sustainability issues that are materially important throughout a product's supply chain and lifecycle (TSC 2021).

A central element of Walmart's sustainability drive was the promotion of its sustainability index which provided suppliers with key performance indicators. According to TSC, "the index helps retailers and suppliers measure sustainability performance for 115 different consumer goods categories. The index not only improves the sustainability of the everyday products we use but also drives efficiency and reduces waste for producers all over the world. TSC creates [key performance indicators] that are used to build category scorecards, which are used to set baselines, benchmark suppliers, track progress, and identify successes and opportunities for improvement" (TSC 2021).

Walmart's approach to improving sustainability outcomes reflected a new step in customer–supplier relations. From the late 1990s, large companies such as automobile or aircraft manufacturers began to require suppliers in the automobile or aviation sector to achieve certification to international standards on environmental management. Their recognition of ISO 14001 certification streamlined a process which previously had required suppliers to answer separate questionnaires from each of their major customers. Answering multiple surveys from Ford, Renault, or Boeing that could amount in the aggregate to hundreds of questions required a significant investment of management time and did not necessarily improve the supplier's environmental performance. Certification of the supplier's management system to ISO 14001 thus achieved the risk reduction objectives the large companies desired and set the certified suppliers on a path toward continual improvement of their environmental management systems.

4.4 Emission Reductions and Removal Enhancements at the Project Level

The establishment of carbon markets to satisfy the demand for flexible mechanisms to meet the declining emissions targets of the 1997 Kyoto Protocol resulted in the development of a variety of mitigation projects. We described in Chap. 2 how GHG mitigation methodologies established by the Clean Development Mechanism (CDM) enabled Annex I parties to the Kyoto Protocol to offset a portion of the GHG emissions they were unable to abate. Each instance of a project type was validated by an independent third party to ensure its eligibility. Following project implementation, emission reductions were verified to ensure that carbon credits were conservatively quantified. The intent of validation and verification was to give confidence to purchasers that "Certified Emission Reductions" (CERs) had truly reduced carbon emissions at the project's location. Issuance of CERs by the CDM Executive Board created a financial instrument that could be traded to investors or parties with an obligation to offset their other emissions.

Standardized project methodologies prescribe the means by which project activities reduce emissions or enhance removals compared to an established baseline. By applying an approved methodology, a project can generate carbon credits issued by an authority or registry. These become financial instruments that can be bought and sold in carbon markets. Mitigation projects come in many types and in nearly every economic sector. They either reduce GHG emissions from an established baseline or enhance GHG removals. Projects report their results in metric tons of CO_2e.

As shown in Fig. 4.6, five economic sectors contribute most to the world's annual GHG emissions. A transition to a low-carbon economy can only succeed if economic actors in these sectors do their parts to reduce emissions.

4.4.1 Multiple Sectors of the CDM

The CDM was created by the climate change secretariat of the UNFCCC to recognize mitigation projects developed in countries not subject to emission limits under the Kyoto Protocol. Industrialized countries, subject to the Kyoto Protocol's emission caps, earned credits that they could use to meet their own emissions targets. Project host countries benefited from the transfer of advanced technologies. Between 2001, which was the first year CDM projects could be registered, and September 7, 2012, the CDM issued 1 billion CERs (CDM 2012a).

The Executive Board approved methodologies for both large-scale and small-scale project activities. Large-scale methodologies could be used for project activities of any size, whereas small-scale methodologies could only be applied if the project activity generated emission reductions below a defined limit.

Beginning in 2001, the CDM Executive Board approved hundreds of methodologies in 15 broad categories (CDM 2006):

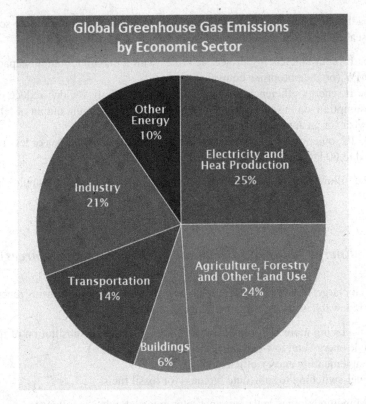

Fig. 4.6 GHG emissions per sector (US EPA 2020)

1. Energy industries (renewable/non-renewable sources)
2. Energy distribution
3. Energy demand
4. Manufacturing industries
5. Chemical industry
6. Construction
7. Transport
8. Mining/mineral production
9. Metal production
10. Fugitive emissions from fuels (solid, oil and gas)
11. Fugitive emissions from production and consumption of halocarbons and sulfur hexafluoride
12. Solvents use
13. Waste handling and disposal
14. Afforestation and reforestation
15. Agriculture.

Broadly speaking, the methodologies were subdivided as either large-scale or small-scale. Small-scale methodologies comprised three different types:

- Type I: renewable energy project activities with a maximum output capacity of 15 MW (or an appropriate equivalent)
- Type II: energy efficiency improvement project activities which reduce energy consumption, on the supply or demand side, with a maximum output of 60 GWh per year (or an appropriate equivalent)
- Type III: other project activities that result in emission reductions of less than or equal to 60 $ktCO_2e$ per year.

In the following pages, we present a small sample of methodologies published by the CDM.

4.4.2 Energy Industries (Renewable/Non-renewable Sources)

This sector targets GHG emission reductions in the energy industries using renewable or non-renewable sources using one of three methods:

(1) Displacing more GHG-intensive technologies through substitution of renewable energy and low-carbon electricity
(2) Implementing energy efficiency
(3) Fuel-switching to substitute biomass for fossil fuels.

CDM methodologies in the energy sector are subdivided as follows:

- Electricity generation and supply
- Energy for industries
- Energy (fuel) for transport
- Energy for households and buildings.

Projects in the first category include replacing electricity generation that use GHG-intensive fuels, such as coal, with renewable energy sources such as wind and solar.

Some CDM methodologies bridge sector boundaries. As an example, AM0053 "Biogenic methane injection to a natural gas distribution grid" is attributed both to the renewable energy sector 1 and the chemical industry sector 5 (CDM 2012b). This methodology is applied to projects that recover, process, and upgrade biogas that is injected into a pipeline for distribution with commercial natural gas. Biogas is recovered from wastewater treatment systems or animal waste management systems and then upgraded to pipeline quality. Only biogas that was previously vented or flared is eligible for crediting under this methodology.

4.4.3 Energy Efficiency

Energy efficiency projects reduce demand for electricity and enable generators to serve more customers with the same installed capacity. Energy efficiency can be obtained by reducing losses and increasing primary energy intensity, indicators used to track progress on global energy efficiency. Technological change and advances in energy management in the industrial and buildings sectors also deliver efficiency improvements. These strategies reduce GHG emissions by decreasing fossil fuel combustion at the power plant (direct emissions) and consumption of electricity by end users (indirect emissions).

The world needs more energy efficiency. The International Energy Agency (IEA) measures energy intensity in terms of tons of oil equivalent per $1000 of GDP (IEA 2020). In 2018, primary energy intensity—an important indicator of how much energy is used by the global economy—improved by just 1.2%, the slowest rate since 2010. This was significantly slower than the 1.7% improvement in 2017 and marked the third year in a row the rate had declined. It was also well below the average 3% improvement goal set in the International Energy Agency's (IEA) Efficient World Strategy, first described in Energy Efficiency 2018 (IEA 2018). According to IEA (2020), the world needs to achieve 2.6% energy efficiency improvements each year until 2030 to meet the UN Sustainable Development Goal (SDG) 7.3 target of "access to affordable, reliable, sustainable, and modern energy for all." The IEA sees electrification of the transportation sector and stricter building codes as key drivers of energy intensity improvements.

4.4.4 Fuel/Feedstock Switching

CDM methodologies enable such industrial sectors as cement, cement blending, inorganic chemical manufacturing, paper, metal, plastic, and others to earn CERs by substituting renewable energy for fossil fuel-based sources. For example, industries making steam from coal-fired boilers could switch to sustainably harvested biomass fuels.

4.4.5 Small-Scale Energy Distribution

This sector seeks to achieve two objectives:

(1) Create the technical conditions to expand access to the electricity network
(2) Repair and maintain existing networks.

A methodology for rural electrification illustrates how a small-scale CDM project may be developed. Methodology AMS.III.AW targets extension of a national or regional grid to rural communities not previously served by electricity. Electricity must come from existing renewable generating plants that extend their power distribution network to the project area. In keeping with the small-scale status of the methodology, emission reductions are limited to 60 $ktCO_2e$ annually (CDM 2005a).

4.4.6 Manufacturing Industries

Manufacturing is one of the largest contributors to CO_2 emissions, but its products can provide technologies that contribute to GHG emissions reductions in other sectors of the economy. The CDM recognizes mitigation measures in this sector as enabling activities since they can support the transition of high-emitting manufacturing sectors to alternative technologies that are consistent with a low-carbon economy (CDM 2019a).

CDM methodologies address several manufacturing-specific project types including those that destroy high-global warming potential chemicals and reduce GHG emissions through specific process changes.

For example, AM0096 addresses the abatement of carbon tetrafluoride (CF_4) in the semiconductor manufacturing sector (CDM 2011a). In this methodology, CF_4 that was previously vented to the atmosphere after being used in the semiconductor etching process is recovered and destroyed in a catalytic oxidation unit (abatement system) located after the etching unit. Methodology AM0065 generates CERs in the magnesium melting industry for facilities that replace sulfur hexafluoride (SF_6) as a cover gas with alternative cover gases having a low or zero global warming potential. Cover gases prevent molten magnesium from oxidizing on contact with air during the scrap remelting and die-cast manufacturing processes (CDM 2008).

4.4.7 Metal Production

Metal production is a highly energy intensive industrial sector contributing to direct GHG emissions in the steel and iron industry and energy indirect (electricity) emissions in aluminum smelters. AM0059, "Reduction in GHGs emission from primary aluminum smelters," applies to project activities that implement technological or operational improvements in primary aluminum production. The project could, for example, upgrade the smelting technology which results in reduction of perfluorocarbon (PFC) emissions or improve the efficiency of electrical energy use (CDM 2016).

4.4.8 Fugitive Emissions from Fuels and Other Substances

CDM methodology AM0009 (version 7.0) addresses the "Recovery and utilization of gas from oil fields that would otherwise be flared or vented" (CDM 2013a). In recognition of the problem of wasted and greenhouse gas emitting flaring, the World Bank established a Global Gas Flaring Reduction Partnership that works with the oil and gas industry around the world to reduce venting and flaring of natural gas (World Bank 2021).

Existing and abandoned coal mines represent a large source of fugitive methane emissions. ACM0008 addresses this issue with a large-scale methodology. The methodology applies when methane is captured and destroyed from coal beds or from mine ventilation systems in new, existing or abandoned coal mines. Under the methodology, all methane captured by the project must either be destroyed or used for power generation, heat or injection into a pipeline. It does not apply to the capture or use of virgin coal bed methane, such as where methane is extracted from coal seams for which there is no valid coal mining concession, or where methane is extracted from abandoned mines that are flooded in accordance with regulations (CDM 2014).

The Montreal Protocol, negotiated in 1986, enshrined agreements from countries around the world to phase out consumption and production of chlorinated refrigerants that scientists had identified as responsible for causing a hole over Antarctica in the atmosphere's protective ozone layer. Chlorinated fluorocarbons such as CFC-11 and CFC-12 were initially replaced by hydrofluorocarbons (HFCs). This helped heal the ozone hole over Antarctica but exacerbated the greenhouse gas problem as HFCs typically had global warming potentials several thousands of times higher than carbon dioxide. HFCs and another class of fluorocarbons, PFCs, were included as classes of engineered gases in the Kyoto Protocol in 1997.

The relevance to climate change of high-GWP hydrofluorocarbons led the CDM Executive Board to approve AM0001 (Version 6.0) "Decomposition of fluoroform (HFC-23) waste streams" to encourage the destruction of powerful greenhouse gases such as HFC-23 (CDM 2011b). The methodology also applied to solvents. Its provisions included the implementation of best practices to reduce fugitive emissions during storage, transfer and transport of HFCs and replacing this class of chemicals with substitutes that reduced the emissions of both greenhouse gases and other volatile organic substances from paints and adhesives.

4.4.9 Buildings and Construction

According to the International Energy Agency (IEA), the building and construction sector globally in 2018 accounted for 36% of final energy use and 39% of energy- and process-related emissions. These levels of GHG emissions placed the sector well ahead of those occupying the second and third places: industry (31%) and transportation (23%). After four years of stable emissions (2013–2016), global emissions from buildings in 2018 increased to 9.7 gigatons (Gt) of CO_2. Population growth contributed to a 1% increase in energy consumption to around 125 exajoules (EJ), or 36% of global energy use (IEA 2019). The importance of this sector was not lost to the CDM which developed methodologies to reduce carbon emissions from construction materials.

AMS-III.BH (CDM 2013b) addresses displacement of production of brick and cement by manufacture and installation of gypsum concrete wall panels. These are less carbon intensive than brick and cement and can be used for non-load-bearing walls, load-bearing walls and fencing (compound/security walls) in green-field building projects or expansion of existing buildings. The methodology limits the use of cement imported to the host country to no more than 10%. This methodology addresses just a small portion of construction activities.

In parallel to methodologies for earning carbon credits for emission reductions, existing standards for certification of energy efficient buildings offer valuable guidance to building planners and developers. The US Green Building Council's LEED program (US GBC 2021) and the BRE Group's BREEAM program (BRE Group 2021) provide standards and certification for environmentally sustainable buildings. Addressing climate change mitigation in buildings is particularly important because the life cycle of a commercial building is around 30 years and around 50 years for housing, meaning that energy inefficiency can be locked into new construction for decades. A modest additional investment at time of construction can save substantial operating expenses over a building's useful life.

A taxonomy under development for use with standards associated with green debt instruments (ISO 2021a, b),[7] observed in the context of green bonds and green loans the importance of energy use during the operational life cycle of buildings: "Construction of energy and resource efficient and low-GHG emission new buildings can make a substantial contribution to climate change mitigation by reducing net GHG emissions from the operational and construction phase of the building life cycle. This should be measured by appropriate indicators of primary energy and net GHG emissions both in the operational phase and along the life cycle (including embodied emissions)" (ISO 2021c).

The ISO taxonomy takes a transitional approach by relying on requirements set in current policies but with a vision to develop and start using, as soon as possible, absolute thresholds for energy and carbon performance. Future thresholds should be based on ambitious performance benchmarks set by building type. The aim will be to set criteria that are at least as ambitious as the level of performance of the top 15% of the local building stock, with the intent to progressively decline to net-zero energy use and net-zero GHG emissions by 2050.

Metrics. At present, mandatory metrics for building construction are found in applicable regulations and building codes, if at all. These may use operational primary energy demand for the assessment of building performance. There is, however, strong

---

[7] ISO/DIS2 14030 Part 3 is under development in 2021 with projected publication in 2022 (ISO 2020).

consensus around the need for the sector to move toward operational net GHG emissions and eventually life cycle operational net GHG emissions (including embodied net GHG emissions). Benchmarks for high performing new buildings cannot be established because performance levels are largely dependent on building type and local climatic conditions.

When establishing site-specific benchmarks, the following performance metrics should be considered:

- Operational primary energy:[8] the annual net primary energy demand during the operational phase of the building life cycle, expressed as kWh/m^2 per year
- Operational net GHG emissions: the annual net carbon-equivalent emission rate (using a GWP over a 100-year time horizon) arising from energy consumption during the operational phase of the building life cycle expressed as $kgCO_2eq/m^2$ per year
- Embodied net GHG emissions: net GHG emissions embodied in building materials during production, transportation and construction and end of life, expressed as $kgCO_2eq/m^2$
- High efficiency walls (U-value better than 1.5 w/m^2K)
- High albedo and solar reflectivity strategies
- Extensive and intensive green roofs.

Thresholds. Absolute thresholds (for operational primary energy as well as operational and embodied net GHG emissions metrics) should be set, as a minimum, at the level of performance corresponding to the top 15% of the local stock. For setting such thresholds, it is necessary to distinguish among building types (at least between residential and non-residential ones) and account for climatic differences. Once established, thresholds for operational net GHG emissions should be projected to decline over time and reach zero emissions by 2050, thus providing a dynamic target as well as a clear indication to the market.

Embodied net GHG emissions. For highly efficient new buildings, net GHG emissions embodied in building materials and in construction and demolition processes, technically referred to in some standards as "embodied GWP" or "carbon emissions," can represent a significant share of the total carbon emitted throughout the building life cycle. The use of a building bill of materials (kg) was considered as a proxy, but it was felt that it does not strongly enough correlate with embodied carbon or reflect possible choices for less GHG emission-intensive building materials. The use of the unit $kgCO_2e/kg$ of product (e.g., concrete, steel or glass) is only approximate at the level of the building. It does not represent all the choices of materials made in the design and realization, because some of them have a variable carbon footprint depending on their density, especially since the carbon impacts according to the materials in the phase of use of the building may be different. International standard methodologies to assess building life cycle emissions exist. Environmental product declarations (EPDs) provide figures for net GHG emissions embodied in

---

[8] Primary energy is energy contained in raw fuels, and other forms of energy received as input to a system. Primary energy can be non-renewable or renewable.

building materials based on life cycle assessment (LCA) and can be combined to produce whole building assessment. However, differences in assessment methods and output formats among EPD issuers pose a significant limitation to the reliability and usability of these certificates.

4.4.10 Transport

Globally, transportation accounts for 14% of greenhouse gas emissions and 24% of CO_2 emissions. Transportation is the largest source of GHG emissions in the USA, surpassing electricity generation in 2016 for the first time. The USA emits more than twice as much CO_2e from transportation as China, the second largest emitter. Europe, whose per capita income is more comparable to that of the USA, is a close third to China in emissions from this sector (Wang and Ge 2019). The difference in emission rates can be explained by a larger use of public transportation in Europe than in the USA.

A pertinent CDM methodology that addresses transportation emissions is AM0031, Bus Rapid Transit (BRT). This methodology applies to project activities that reduce emissions through the construction and operation of a new BRT system or lane(s) for urban road-based transport, or expansion of existing BRT systems. A BRT system can supplement an urban rail-based rapid transit system but not replace it. The goal of the methodology is to displace more GHG-intensive transportation modes by less GHG-intensive ones (CDM 2019b).

4.4.11 Waste Handling and Disposal

Waste management is a sector with multifaceted environmental aspects. Regulations in developed countries seek to control litter and to prevent the contamination of stormwater runoff. Waste reduction campaigns through training and public outreach encourage reduction in the volume of wastes sent to landfills through materials reuse and recycling. Land for urban waste management is increasingly scarce or distant and heightened environmental awareness has incentivized industry and the public to recycle more. These values and the cost of waste management have combined to reduce the number of operating landfills despite growth in population and consumption.

In the USA, residents in 2018 generated 292.4 million tons (US short tons) or 4.9 pounds per person per day of municipal solid waste (MSW). Of this total, approximately 69 million tons were recycled, and 25 million tons were composted, for a 32% recycling and composting rate. About 146.2 million tons of MSW were landfilled. The largest component of landfilled waste was food at about 24%. Plastics were second at 18%, paper and paperboard third at about 12%, and rubber, leather, and textiles comprised about 11%. Other types of materials accounted for less than 10% each (US EPA 2021b).

Landfill gas (LFG) is produced from the decomposition of organic material in landfills. It is composed of approximately 50% methane, 50% carbon dioxide (CO_2), and a small amount of non-methane organic compounds. Current US regulations require landfill gas collection and combustion at some, but not all, MSW landfills (US EPA 2021b).

CDM methodology ACM0001 "Flaring or use of landfill gas" (version 19.0) encourages the collection and combustion of landfill gas. The collected gas can be used to generate electricity, provide heat, be upgraded to pipeline quality natural gas, or flared (CDM 2019c).

4.4.12 Agriculture

Greenhouse gas emissions from the Agriculture, Forestry, and Other Land-Use (AFOLU) sector amount to nearly one quarter of all global emissions. With a share of 24% in 2020, global AFOLU emissions were only slightly outdistanced by the electricity and heat generation sector which accounted for 25% (EPA 2020a). Outputs from the agriculture sector play an important role in nourishing the world's populations and providing employment to many. The sector offers many mitigation opportunities as well as ways to make our economies more circular and therefore less wasteful.

Rice cultivation. Rice cultivation provides a staple crop that feeds billions of people. The crop is grown in many parts of the world and in many climates but prospers in tropical regions. Rice grows best in standing water and for that reason is a major contributor to climate change. When water in rice paddies become anaerobic, bacteria generate methane (Nguyen 2004). Rice cultivated in low-lying regions is also susceptible to the effects of climate change due to sea-level rise. According to Klutger and Lemonick, cited by Nguyen (2004 p 26), coastal areas where much rice is cultivated will become uninhabitable once sea levels have risen by as much as 88 cm. Locations of affected lands range from Bangladesh in Asia to Louisiana in the Gulf of Mexico.

In the USA, the Climate Action Reserve developed a rice cultivation protocol to encourage the reduction of methane emissions. Eligible projects do this by engaging in dry seeding of rice which delays field flooding and reduces methane emissions. Another eligible activity is the removal of post-harvest rice straw from the field. This management activity lowers methane emissions from the effects of decomposition of rice straw. The protocol is applicable in California only (Climate Action Reserve 2013, pp 6–7).

Soil carbon enhancement. Productive soils include a healthy proportion of soil carbon and other organic matter. Soil organic carbon (SOC) improves the structure of soil and raises crop yields. Soil is naturally a carbon sink and "sequesters" carbon. Several agricultural practices enhance SOC sequestration (Corning 2016):

- The use of conservation tillage, where routine turning over of soil with a plow is replaced by no-till or low-till planting practices
- Crop residue management, where the stalks of harvested plants are left on the ground, so they decompose and add carbon to the soil
- Using cover crops to add biomass both above ground and in root structures
- Applying organic amendments to the soil such as manure and compost
- Planting perennial crops which increase SOC through post-harvest decomposition of both roots and litter.

To help incentivize soil carbon enrichment practices, the Verified Carbon Standard in 2012 issued its VM0021 Soil Carbon Quantification Methodology developed by The Earth Partners LLC (Verified Carbon Standard 2012). A Climate Action Reserve Soil Enrichment Protocol in 2020 similarly targeted agricultural practices designed to increase carbon stocks in soil (Climate Action Reserve 2020).

Afforestation and Reforestation. Forests have been called the lungs of the planet because they absorb CO_2 from the atmosphere. Since the adoption of the Kyoto Protocol, projects to enhance the removal of CO_2 from forests have been advocated for the following reasons:

- To offset rising CO_2 emissions in other sectors
- To protect the Amazon and other forests from deforestation
- To enhance biodiversity
- To protect the livelihoods of indigenous peoples living in the forest.

Although the CDM recognized afforestation and reforestation project methodologies (CDM 2005b), the sector contributed very few removal enhancements in the Kyoto Protocol's first commitment period as CERs were issued to only one project among 40 that were registered (Carbon Market Watch 2012). The main activity in forest projects migrated to the voluntary carbon markets through programs operated by ACR, VCS, and the World Bank. Forest carbon credits did not become a mainstay of regulated cap-and-trade programs until 2011 when the state of California recognized three types of forest projects that could be implemented in any US state except Hawaii (CARB 2015 p 25).

California's approved project types include:

- Reforestation: restoration of tree cover on land that at the time of project implementation had no, or minimal, tree cover
- Improved forest management: activities that increase carbon stocks on forested land relative to baseline levels of carbon stocks
- Avoided conversion: specific actions that prevent the conversion of privately owned forestland to a non-forest land use by dedicating the land to continuous forest cover through a conservation easement or transfer to public ownership.

California issued Air Resources Board Offset Credits (ARBOCs) in metric tons of CO_2 based on the amount of forest carbon stocks that are additional to the baseline during each reporting period. The program includes a buffer account to which each project contributes based on an analysis of reversal risks. Reversals may be either unintentional, such as by wildfires, or intentional, such as by timber harvesting. Unintentional reversals result in a one-to-one replacement of the reversed carbon stock offset credits from the buffer account. Intentional reversals require replacement of offset credits by the project developer. If the project type is improved forest management and the project duration is less than fifty years at the time of the reversal, the project developer is required to offset the intentional reversal with a greater number of compliance instruments than those initially earned (CARB 2015: 33–34).

4.5 Critique of a Durban (SA) MSW CDM Project

This chapter has provided an overview of the types of methodologies and protocols developed during the first two decades of the twenty-first century to help mitigate greenhouse gas emissions. In Chap. 2, we discussed limitations of carbon trading as a policy instrument to achieve decarbonization of economies. To be sure, the generation and use of carbon credits have been controversial since their first implementation after the Kyoto Protocol entered into effect in 2005. The run-up to COP21 in Paris occasioned numerous conferences, including one held on November 9, 2015, by the Veolia Institute.[9] In a session focused on the problem of fugitive methane emissions, John Parkin, Deputy Director of the Facilities and Engineering Division of the Cleanup and Solid Waste Department (DSW) of the Municipality of Ethekwini (Durban), discussed his experience as the host of a CDM landfill gas project. Key points from his experience included:

1. The economic case for the project was based on 50% of revenues derived from consumers of the landfill-generated electricity and 50% by the sale of CERs. However, the project had not anticipated falling CER prices. In fact, the price of CERs peaked somewhere in 2009–10 and from then gradually decreased. In 2012, CERs were worth around $1.00/tCO$_2$e, and by 2015, when the validity of pre-2012 CERs expired, CER prices crashed to $0.10-$0.20 (Hemlata 2019).
2. The project required an investment of about €15 million over 11 years from the appointment of the World Bank as a project partner to the delivery of the first kWh to neighborhood residents.
3. CDM project managers lacked technical competence on methane capture, and this resulted in a large turnover of staff.
4. The project incurred higher than expected costs including payments to advisors and auditors and reimbursement of their travel costs.

Publicly traded Veolia was the lead project manager in this project. Like many waste professionals, Veolia for many years had considered methane from landfills to be an untapped resource. They had not anticipated the difficulties associated with managing this project to completion on time and within budget.

---

[9] A think-tank sponsored by the Veolia Foundation (Veolia 2021).

4.6   Issues Associated with the CDM

It took time for EU ETS market participants to discover the proper value of CDM's CERs. It took even longer for ERUs from Joint Implementation to be introduced into the market. As discussed in Chap. 2, the market's response to the first signs that European Union Allowances (EUAs) had been allocated to industry in excess of their compliance needs severely reduced demand for CERs.

1. Early CER prices rose to close to $40/tCO$_2$ before falling to around $10 in 2008. The quantity of CER issuances grew through 2012 despite a declining carbon price which then collapsed in 2012 at the end of the first Kyoto commitment period (2012).
2. A relatively low volume of carbon credits were issued in voluntary offset markets with relatively low credit prices peaking in 2008 and steadily declining thereafter.
3. The EU ETS disallowed the further use of credits from some projects (e.g., HFC-23 reduction, largely in China) in the second Kyoto commitment period (2013–2020), as the project type appeared to incentivize over-production of the HCFC refrigerant R22.

A significantly high proportion of CDM projects (3,414 out of 11,262, or 30%) were initiated but later abandoned (CDM 2021). O. Chima Okereke (2017) analyzed eight cases of project failure in Africa and traced the main root causes to:

– Corruption (due to a lack of governance)
– A lack of skills, absence of training, and ignorance of project management
– A lack of resources, failure to include the local communities in planning, project implementation, operations and maintenance management, and no arrangements made for ensuring the delivery of project outputs.

Centralized decision-making and management oversight by the CDM Executive Board contributed to some of the outcomes associated with the Durban case study. These included high project development costs and extended timeframes for issuing credits. Another factor, the overallocation of EUAs by the European Union, contributed to the collapse in price of CERs and upended the financial projections of many project developers.

The future of the CDM has been in doubt since its role under the Kyoto Protocol came to an end in 2020. Negotiators at successive COPs (COP24 in Poland in 2018, COP25 in Spain in 2019) failed to agree on the rules and modalities for implementing Article 6 of the Paris Agreement. Paragraph 4 of this article provides for the creation of a mechanism to contribute to the mitigation of greenhouse gas emissions. This mechanism could be a redefined CDM or a newly constituted entity. Proponents of the use of carbon markets to mitigate emissions have expressed hope that decisions on how to implement this Article could be reached in Glasgow at the COP26 in November 2021.

4.7 Voluntary Carbon Offset Programs

A chapter on mitigation projects would be incomplete without recognizing programs other than the CDM. The VCS enabled project developers to use project methodologies approved by the CDM Executive Board anywhere in the world. Meanwhile, other organizations, such as the American Carbon Registry and the Climate Action Reserve in the USA, launched voluntary project protocols and established trading mechanisms to attract investment to carbon mitigation projects.

In 2007, a group of carbon offset service providers established the International Carbon Reduction and Offset Alliance (ICROA, now part of IETA) to promote voluntary action as a valuable complement to compliance and regulation. Members' compliance to a Code of Best Practice is audited by a third-party annually. This self-regulatory initiative brought an overarching approach to quality assurance for buyers of offset-inclusive carbon management services, as it defined accepted practices for measuring, reducing, and offsetting emissions, and communicating climate actions accurately (Shopley 2015). ICROA's efforts have not prevented all abusive practices in carbon markets, particularly on unregulated trading platforms where transactions in carbon credits suffer from poor pricing transparency and at times profiteering by traders (Redshaw 2021).

Concerns about the quality of voluntary carbon credits prompted the NGO-backed Gold Standard in 2015 to introduce requirements for the demonstration of sustainable development aspects in CDM and voluntary carbon projects. This enhanced project type has attracted investors to small-scale projects and projects in least-developed economies and increased access in communities to carbon finance (Shopley 2015).

Voluntary carbon markets in North America have been largely sustained by the desire of businesses to offset their emissions of GHGs to demonstrate action to combat climate change. Those in regulated markets have purchased carbon credits to stay within caps on emissions set by state environmental authorities. Since 2005, the principal voluntary carbon market issuers[10] together have issued approximately 1.3 billion ERTs, CRTs, and VCUs each representing a ton of CO_2e emission reductions or removal enhancements. Annual issuances in 2019 exceeded 100 million metric tons of CO_2e (Ecosystem Marketplace 2020, p 2).

In 2020, a task force led by Tim Adams, Bill Winters, and Annette Nazareth—supported by Mark Carney[11]—announced an initiative to "scale" the voluntary carbon market in anticipation of significantly increased demand for offset credits by businesses during the 2020s. The task force issued its final report in January 2021 and concluded that voluntary carbon markets would need to increase available credits

---

[10] The main sources of voluntary carbon credits over this period are the American Carbon Registry, the Chicago Climate Exchange, the Climate Action Reserve, and Verra, the parent organization for the VCS.

[11] The Task Force leaders are, respectively, president and CEO of the Institute of International Finance, CEO of Standard Chartered, Senior Counsel at Davis Polk and former commissioner of the US Securities and Exchange Commission. Mark Carney is UN Special Envoy for Climate Action and Finance and Finance Advisor to UK Prime Minister Boris Johnson for COP26.

by a factor of 15 by 2030. The final report also highlighted the need to ensure the quality of issued credits and proposed that a new mechanism be created to ensure adherence to quality principles (IIF 2021).

A steady increase in voluntary carbon emission reductions and removal enhancements can positively augment greenhouse gas mitigation efforts as long as carbon credits are real, additional, quantifiable, permanent, verifiable, and enforceable. To play a meaningful role in the transition to a low-carbon economy, these carbon credits also must not serve as a substitute for deep decarbonization of hard-to-abate industries. Climate actions that fail to reduce emissions on the scale and at the pace needed to achieve net-zero emissions by 2050 represent at best a palliative where surgery and innovation are required; at worst, they represent a strategy to divert attention from the transformative measures needed to protect planet Earth from increasingly catastrophic warming.

Questions for Readers

1. Why is the concept of additionality easier to understand than apply?
2. Why do investors and others insist that mitigation projects be independently validated and verified?
3. Why is a ton of CO_2 removed from the atmosphere equivalent to a ton of CO_2 emission reductions?
4. What is the difference between a carbon credit issued under regulatory and voluntary GHG programs?
5. How do GHG programs take into consideration environmental objectives other than GHG emissions and removals?

References

Boyer Kind E (2018) Trial of carbon tax 'fraud of the century' opens in Paris. France 24. Broadcast on: 29/01/2018–18:19. https://www.france24.com/en/20180129-france-trial-carbon-credits-fraud-paris-crime-emissions-scam-melgrani-marseille. Accessed on 15 June 2021

BRE Group (2021) https://www.bregroup.com/. Accessed on 15 May 2021

CARB (2015) Compliance offset protocol. U.S. forest projects. https://ww2.arb.ca.gov/sites/default/files/classic//cc/capandtrade/protocols/usforest/forestprotocol2015.pdf. Accessed 19 Feb 2021

Carbon Market Watch (2012) Forestry/Land use projects in the CDM. https://carbonmarketwatch.org/2012/05/30/forestry-projects/. Accessed 19 Feb 2021

CDM (2005a) AMS-III.AW.: electrification of rural communities by grid extension—version 1.0. https://cdm.unfccc.int/methodologies/DB/GRH88B4S68PO9H0YELQ8ZMVANO14JR. Accessed 12 June 2021

CDM (2005b) Simplified baseline and monitoring methodologies for selected small-scale afforestation and reforestation project activities under the clean development mechanism. Version 1. https://cdm.unfccc.int/methodologies/DB/91OLF4XK2MEDIRIWUQ22X3ZQAOPBWY. Accessed 5 Feb 2021

CDM (2006) List of sectoral scopes. https://cdm.unfccc.int/DOE/scopelst.pdf. Accessed 13 Nov 2020

CDM (2008) AM0065: replacement of SF6 with alternate cover gas in the magnesium industry—version 2.1. https://cdm.unfccc.int/methodologies/DB/GNX2U6RAUIP1UD1IP3CRDPVPPIGSS0. Accessed 12 June 2021

CDM (2011a) AM0096: CF4 emission reduction from installation of an abatement system in a semiconductor manufacturing facility—version 1.0.0. https://cdm.unfccc.int/methodologies/DB/SF95S0OW4343SA06Z6FUUXYD0FFTCT. Accessed 12 June 2021

CDM (2011b) AM0001: decomposition of fluoroform (HFC-23) waste streams—version 6.0.0. https://cdm.unfccc.int/methodologies/DB/GAOZAY2DWIQHK71LJS027N6N4AV6SC. Accessed 12 June 2021

CDM (2012a) News release: Kyoto protocol's CDM passes one billionth certified emission reduction milestone. https://cdm.unfccc.int/CDMNews/issues/issues/I_P0QZOY6FWYYKFKOSAZ5GYH2250DRQK/viewnewsitem.html. Accessed 16 Nov 2020

CDM (2012b) AM0053: biogenic methane injection to a natural gas distribution grid—version 4.0.0. https://cdm.unfccc.int/methodologies/DB/FKDGZEEEQC4XNUT326116FS0S8USP1. Accessed 12 June 2021

CDM (2013a) AM0009: recovery and utilization of gas from oil fields that would otherwise be flared or vented—version 7.0. https://cdm.unfccc.int/methodologies/DB/ET4NXMVXFQ5C2EJ5L1OF8YZIEVLVDA. Accessed 12 June 2021

CDM (2013b) AMS-III.BH.: displacement of production of brick and cement by manufacture and installation of gypsum concrete wall panels—version 1.0. https://cdm.unfccc.int/methodologies/DB/YZBSIH9BCH894GDSD4BP2FMNMI9FU6. Accessed 12 June 2021

CDM (2014) ACM0008: abatement of methane from coal mines—version 8.0. https://cdm.unfccc.int/methodologies/DB/YSD3FQ5WR3VPC9Q64CDTLXHLFVKKKU. Accessed 12 June 2021

CDM (2016) AM0059: reduction in GHGs emission from primary aluminium smelters—version 2.0. https://cdm.unfccc.int/methodologies/DB/CHNLRVLNEAM438MR5400YQDS3CPC50. Accessed 12 June 2021

CDM (2019a) CDM methodology booklet. https://cdm.unfccc.int/methodologies/documentation/meth_booklet.pdf. Accessed 13 Nov 2020

CDM (2019b) AM0031. Large-scale methodology. Bus rapid transit projects, version 7.0. https://cdm.unfccc.int/methodologies/DB/7DF4Q82IMUANFW97FIUFRHBZQTKXVF. Accessed 27 Feb 2021

CDM (2019c) ACM0001: flaring or use of landfill gas—version 19.0. https://cdm.unfccc.int/methodologies/DB/JPYB4DYQUXQPZLBDVPHA87479EMY9M. Accessed 12 June 2021

CDM (2021) Clean development mechanism: CDM projects. https://cdm.unfccc.int/index.html. Accessed 6 June 2021

Climate Action Reserve (2013) Rice cultivation project protocol. https://www.climateactionreserve.org/how/protocols/rice-cultivation/. Accessed 19 Feb 2021

Climate Action Reserve (2020) Soil enrichment protocol. https://www.climateactionreserve.org/how/protocols/soil-enrichment/. Accessed 19 Feb 2021

Corning et al (2016) The carbon cycle and soil organic carbon. http://nmsp.cals.cornell.edu/publications/factsheets/factsheet91.pdf. Accessed 19 Feb 2021

Ecosystem Marketplace (2020) Forest trends' ecosystem marketplace, voluntary carbon and the post-pandemic recovery. State of voluntary carbon markets report, special climate week NYC 2020 installment. Washington DC: forest trends association. https://www.forest-trends.org/publications/state-of-voluntary-carbon-markets-2020-voluntary-carbon-and-the-post-pandemic-recovery/. Accessed 19 Feb 2021

Hemlata K (2019) Rise and fall of carbon credits: finding solutions for climate crisis in free market (eds) TWC India. https://weather.com/en-IN/india/news/news/2019-11-02-rise-fall-carbon-credits-solutions-climate-crisis-free-market. Accessed 18 March 2021

IEA (2018) Energy efficiency 2018. https://www.iea.org/reports/energy-efficiency-2018. Accessed 15 Nov 2020

IEA (2019) Changes in floor area, population, buildings sector energy use and energy-related emissions globally, 2010–18. https://www.worldgbc.org/sites/default/files/2019%20Global% 20Status%20Report%20for%20Buildings%20and%20Construction.pdf. Accessed 5 Feb 2021

IEA (2020) SDG 7: data and projections. Energy intensity. October 2020. https://www.iea.org/rep orts/sdg7-data-and-projections/energy-intensity. Accessed 27 Feb 2021

IIF (2021) Task force on scaling voluntary carbon markets: final report. Institute of international finance. https://www.iif.com/tsvcm. Accessed 12 June 2021

ISO (2019a) ISO 14064 greenhouse gases—part 2: specification with guidance at the project level for quantification, monitoring and reporting of greenhouse gas emission reductions or removal enhancements. https://www.iso.org/standard/66454.html. Accessed 5 Feb 2021

ISO (2019b) ISO 14064 greenhouse gases—part 3: specification with guidance for the verification and validation of greenhouse gas statements. https://www.iso.org/standard/66455.html. Accessed 23 June 2020

ISO (2021a) ISO 14030-1, environmental performance evaluation—green debt instruments— process for green bonds. https://www.iso.org/standard/43254.html. Accessed 12 June 2021

ISO (2021b) ISO 14030-2, environmental performance evaluation—green debt instruments— process for green loans. https://www.iso.org/standard/75558.html. Accessed 12 June 2021

ISO (2021c) ISO/DIS2 14030-3 environmental performance evaluation—green debt instruments— part 3: Taxonomy. https://www.iso.org/standard/75559.html. Accessed 12 June 2021

Millington D et al (2020). The economic effectiveness of different carbon pricing options to reduce carbon dioxide emissions. Canadian energy research institute, Calgary, AB, Study No. 189, 2020. https://ceri.ca/assets/files/Study_189_Full_Report.pdf. Accessed 1 March 2021

Nguyen N (2004) http://www.fao.org/forestry/15526-03ecb62366f779d1ed45287e698a44d2e.pdf. Accessed 22 June 2020

Okereke OC (2017 Jan) Causes of failure and abandonment of projects in PM. World Journal VI(I)— project deliverables in Africa. https://pmworldlibrary.net/wp-content/uploads/2017/01/pmwj54- Jan2017-Okereke-causes-of-project-failures-in-africa-featured-paper2.pdf. Accessed 20 March 2021

Peteritas B (2013) The California air resources board has linked its program for cutting green- house gas emissions and curbing climate change with one in the Canadian province of Quebec Brian. April 23, 2013. https://www.governing.com/archive/mctcalifornia-cap-and-trade- program-linked-to-quebecs.html. Accessed 16 Feb 2021

Redshaw L (2021) Comment: voluntary carbon market—broken more than breakthrough. In Carbon Pulse 18 June 2021 https://carbon-pulse.com/category/newsletter/. Accessed 18 June 2021

Shopley J (2015) The road to Paris: balancing ambition and commitment. In natural capital partners. 2 June 2015. https://www.naturalcapitalpartners.com/news-resources/post/the-road-to-paris-bal ancing-ambition-commitment. Accessed 26 May 2021

Stainforth T (2020) More than half of all CO2 emissions since 1751 emitted in the last 30 years, Institute of European environmental policy, 29 April 2020. https://ieep.eu/news/more-than-half- of-all-co2-emissions-since-1751-emitted-in-the-last-30-years. Accessed 25 May 2021

Tigray Regional State (2015) Regional REDD+ coordination unit tigray regional state, Mekele September 3/2015. https://slideplayer.com/slide/14438110/ p 17. Accessed 17 Feb 2021

TSC (2019) Case studies: retailer commitments. https://www.sustainabilityconsortium.org/projects/ retailer-commitments/. Accessed 15 Feb 2021

TSC (2021) Case studies: walmart sustainability index. https://www.sustainabilityconsortium.org/ impact/case-studies/walmart-sustainability-index/. Accessed 15 Feb 2021

UN Environment Programme (2020) Emissions gap report 2020. Nairobi. https://www.unenviron ment.org/emissions-gap-report-2020. Accessed 15 Jan 2021

US DOE (2021) Energy service companies. https://www.energy.gov/eere/femp/energy-service-com panies-0. Accessed 15 Feb 2021

US EPA (2020) Global greenhouse gas emissions data. https://www.epa.gov/ghgemissions/global- greenhouse-gas-emissions-data. Accessed 4 Feb 2021

US EPA (2021a) About energy star/EPA's role. https://www.energystar.gov/about/origins_mission/epas_role_energy_star. Accessed 15 Feb 2021

US EPA (2021b) Facts and figures about materials, waste and recycling. https://www.epa.gov/facts-and-figures-about-materials-waste-and-recycling. Accessed 9 Feb 2021

US GBC (2021) U.S. green building council. https://www.usgbc.org/. Accessed 27 May 2021

Veolia (2021) https://www.veolia.com/. Accessed 10 Feb 2021

Verified Carbon Standard (2012) Soil carbon quantification methodology. https://verra.org/wp-content/uploads/2018/03/VM0021-Soil-Carbon-Quantification-Methodology-v1.0.pdf. Accessed 12 June 2021

Walmart (2020) 2020 environmental, social and governance report. https://cdn.corporate.walmart.com/90/0b/22715fd34947927eed86a72c788e/walmart-esg-report-2020.pdf. Accessed 15 Feb 2021

Walmart (2021) Audits, certifications and testing. https://corporate.walmart.com/suppliers/requirements/audits-and-certifications. Accessed 15 Feb 2021

Wang S, Ge M (2019) Everything you need to know about the fastest-growing source of global emissions: transport. world resources institute. 16 October 2019. https://www.wri.org/blog/2019/10/everything-you-need-know-about-fastest-growing-source-global-emissions-transport. Accessed 27 Feb 2021

World Bank (2021) Global gas flaring reduction partnership. https://www.worldbank.org/en/programs/gasflaringreduction. Accessed 16 May 2021

Chapter 5
Measuring, Reporting, and Verification

5.1 Measurement and Monitoring

The title of this chapter—Measuring, Reporting and Verification—is both a term of art in international climate change frameworks and a set of words that describe a logical sequence of steps in quality assurance and quality control of data. The term, often referred to by its abbreviation MRV, took on greater prominence at the 17th Conference of the Parties in Durban, South Africa, in 2011. Developed countries who planned to assist developing countries with the implementation of "nationally appropriate mitigation actions" (NAMAs) stressed the importance of associating actions with complete, accurate, and transparent data reporting (Lütken et al. 2012). MRV remains an important element in international climate change accounting even though NAMAs have been superseded under the Paris Agreement by programs and projects to implement nationally determined contributions (NDCs) submitted to the United Nations Framework Convention on Climate Change (UNFCCC).

Countries designated in Annex I of the Kyoto Protocol—the world's most industrialized countries—report emissions of GHGs to the UNFCCC each year by April 15 (UNFCCC 2020a). Non-Annex I Parties have greater flexibility in reporting. They are required to submit their first national communication, including information on GHG emissions/removals, within three years of entering the convention and every four years thereafter. Updates to the national communications information are provided every two years in a biennial update report (UNFCCC 2020b).

While MRV in the UNFCCC context emphasizes "measurement," the initial "M" may also stand for "monitoring." The distinction is subtle, but important. All values that are measured can be monitored, but not all monitored parameters are capable of measurement. For this reason, it is more common to use the term "monitoring" in mitigation project development. Some parameters are constants, such as global warming potentials; some are assumptions, such as default discounts to account for uncertainty of measurement; and some are aggregated data, such as units of product sold. In this chapter, we discuss both measurement and monitoring, and will use the term appropriate for each context.

© The Author(s), under exclusive license to Springer Nature Switzerland AG 2021   107
J. C. Shideler and J. Hetzel, *Introduction to Climate Change Management*,
Springer Climate, https://doi.org/10.1007/978-3-030-87918-1_5

It is uncommon for greenhouse gas emissions to be measured directly. There are exceptions, such as CO_2 measurement devices in Continuous Emissions Monitoring Systems (CEMS), but these are not widely deployed in industry. In most cases, GHG inventory managers use "activity data" and "emission factors" to quantify greenhouse gas emissions.

5.1.1   Activity Data and Emission Factors

Activity data describe substances or processes that will generate GHG emissions. Liters or cubic meters of fossil fuels are good examples. It is easy to add up the total volume of fossil fuels an organization purchases in a year. But these numbers do not provide us with equivalent tons of CO_2 emissions. For that, the inventory manager needs emission factors. Emission factors convert a stated amount of combusted fuel into carbon dioxide (CO_2) emissions, methane (CH_4) emissions, and nitrous oxide (N_2O) emissions. Global warming potentials (GWPs) then normalize the emissions of methane and nitrous oxide into equivalent amounts of "carbon dioxide equivalent" (CO_2e) emissions. We provide an example for gasoline in Tables 5.1 and 5.2.

Emission factors for the consumption of fossil fuels are fairly reliable as they are based on stoichiometric analyses of the combustion process. Emission factors for other fuels may be less reliable, such as for wood combustion, where the amount of moisture in the wood is an important variable. In any case, most GHG inventories rely on activity data and emission factors for the quantification of GHG emissions.

GHG inventory managers should prefer emission factors that correspond as much as possible to local conditions. Regulatory programs may require oil and gas companies to perform a monthly analysis of the carbon content of the "field gas" that is produced along with oil extracted from oil wells. Emission factors for electricity vary by geography, so it is important that GHG inventory managers obtain emission factors for the average of electricity production plants available in their region. In

Table 5.1 Emission factors for Canadian motor gasoline (Environment and Climate change Canada, (ECCC 2019 Table A6.4, p 222)

Fuel	Unit	g CO_2	g CH_4	g N_2O
Motor gasoline	Liter	2307	0.1	0.02

Table 5.2   Quantification of CO_2e of gasoline using Canadian emission factors

Gasoline	Emission Factor	tGHG	GWP	tCO_2-e/L
CO_2 in g/L	2307	0.002307	1	0.002307
CH_4 in g/L	0.1	0.0000001	28	0.0000028
N_2O in g/L	0.02	0.00000002	265	0.0000053
Total CO_2e/L				0.0023151

the USA, these vary considerably as different regions have varying proportions of generating stations powered by coal, gas, and nuclear fuels along with renewable energy sources such as wind, solar, and hydroelectric.

Where locally calculated emission factors are not available, the IPCC publishes emissions factors that can be used as representative for the power grids of different countries or that represent averages applicable to different economic sectors.

The IPCC is also the main source for GWPs. Kyoto Protocol reporting relied on the IPCC's Second Assessment Report (SAR), which was published in 1996. Its values are still in use for some programs. Most, however, have transitioned their GWP values to either the IPCC's Fourth Assessment Report (AR4) or Fifth Assessment Report (AR5). It is important in any inventory to identify which GWPs are used. Some programs now request that GHGs be quantified in accordance with the "latest" IPCC GWPs.

5.1.2 Monitoring Methods and Equipment

Meters are a common type of monitoring equipment for GHG reporting. Flow meters are used to provide volumetric activity data for liquid and gaseous fuels. Electricity meters measure the amount of power delivered to a customer in an hour. The unit for invoicing purposes is expressed in kilowatt hours (kWh). Fuel and electricity meters that are used in commerce are deemed to be accurate for GHG reporting purposes. Meters installed by organizations for their own use need to be maintained and calibrated on a regular basis to ensure accurate data reporting.

Gaseous fuels other than pipeline quality natural gas may require gas analyzers to measure the amount of methane in the gas. Landfills that capture the gas produced by the decomposition of municipal solid waste produce a landfill gas that is typically 50% methane and 50% carbon dioxide, but the proportion of those two gases varies depending on local conditions. For this reason, it is necessary to know not just the volume of landfill gas produced but also its methane concentration.

Industries that monitor their use of GHGs internally employ instrument technicians to maintain flow meters and gas analyzers and to calibrate them on a regular schedule.

Some organizations store fuel in on-site tanks. Fuel dispensed from such tanks may be metered each time fuel is delivered to a user, and the fuel tickets retained for accounting purposes and to quantify the amount of fuel dispensed over time. It is also possible to calculate the annual amount of fuel consumed by using an inventory accounting method where the difference in tank volumes at the beginning and end of a year are added or subtracted from total additions to the fuel tanks during a year. If the tank level is lower at the end of the year than at the beginning, the difference is added. If the tank level is higher at the end of the year than at the beginning, the difference is subtracted. For greater certainty, an organization may use both methods to account for fuel consumed and employ a process to reconcile any differences.

5.1.3   The IPCC Guidelines for National Reporting

Since the establishment of the United Nations Framework Convention on Climate Change (UNFCCC) in the early 1990s, parties to the convention have submitted national inventories of greenhouse gas emissions on a regular basis. These reports are prepared for their respective governments using a "top-down" approach that gathers and summarizes emissions data from a variety of sources. In the USA, the Environmental Protection Agency (US EPA) publishes information summarizing the federal government's estimates of national emissions (US EPA 2020a).

US EPA has prepared an *Inventory of US Greenhouse Gas Emissions and Sinks* since the early 1990s. This annual report provides a comprehensive accounting of greenhouse gas emissions in the USA. The gases covered by the Inventory include CO_2, CH_4, nitrous oxide (N_2O), hydrofluorocarbons (HFCs), perfluorocarbons (PFCs), sulfur hexafluoride (SF_6), and nitrogen trifluoride (NF_3). The inventory also calculates CO_2 removals from the atmosphere by "sinks," e.g., through the uptake of carbon and storage in forests, vegetation, and soils (US EPA 2020b). Figure 5.1 summarizes this data in graphic form for the year 2018.

5.1.4   Monitoring at the Organizational Level

Monitoring of GHG emissions at the organizational level for many organizations is voluntary. Where monitoring and reporting is required by law, many mandatory programs require organizations to develop and follow monitoring plans whose contents are specified in regulations. Organizations that do not fall in this category must decide on their own whether to monitor GHG emissions and how.

Organizations that have implemented ISO 14001, Environmental management systems—Requirements with guidance for use (ISO 2015), should consider GHG emissions as an "aspect" of their activities, products, or services, and evaluate them for significance. This analysis should take a life cycle perspective into account, so that embodied carbon in the supply chain is considered as well as emissions from end users of the organization's products. If the organization is required by law to report its GHG emissions, this is identified by the organization as a "compliance obligation." Regardless, the organization may determine GHG emissions to be a significant environmental aspect, a condition that triggers many further requirements of the standard.

Users of ISO 14001 are required to plan actions to address significant environmental aspects and compliance obligations, as well as risks and opportunities the organization has identified. Organizations that implement ISO 14001 monitor, measure, analyze, and evaluate their environmental performance. They also conduct internal audits and perform reviews at which top management evaluates the environmental management system to ensure its continuing suitability, adequacy, and effectiveness.

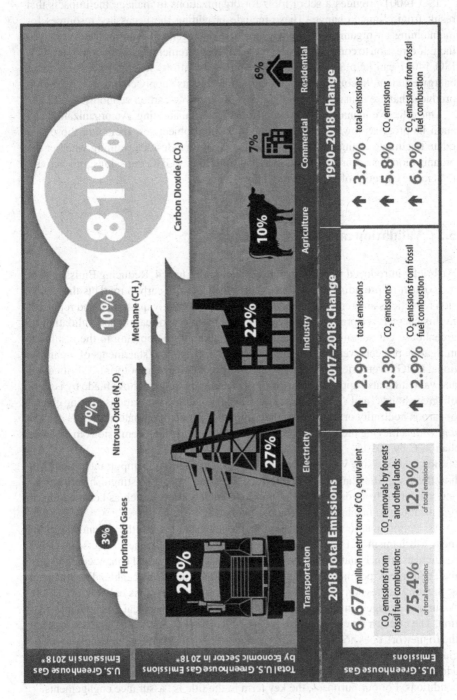

Fig. 5.1 National-level US greenhouse gas inventory (US EPA 2020c)

ISO 14001 provides a set of tools for organizations to manage the impacts that result from climate change. These include providing the necessary resources for maintaining an organization's environmental management system. The tools also include direction to control its processes and its procurement of products and services. ISO 14001 can help an organization mobilize the efforts needed to become more energy efficient, to consider the carbon footprint of its supply chain, and to develop products that are consistent with the transition to a low-carbon economy.

One effective management systems tool is internal auditing. An organization can audit its progress toward meeting its policies and objectives, including those that relate to climate change. If the scope of the audit includes determining whether the organization has accurately quantified and reported its GHG emissions, this activity falls in the category of GHG verification.

5.2   Validation and Verification

We briefly introduced validation and verification in Chap. 4, Reducing Emissions, in the context of providing assurance to participants in the carbon markets that emission reductions and removal enhancements were accurately quantified and reported. In this chapter, we provide further details about how verification and validation of greenhouse gas statements are performed. In addition to applying to the results of mitigation projects, verification and validation also apply to statements of organizational GHG inventories and to statements of product carbon footprints. Verification and validation also apply to other types of statements, such as those made by issuers of "green bonds" and "green loans" (see Chap. 6) and to organizations making claims of carbon neutrality or stating that their implementation of climate actions is consistent with achieving the Paris Agreement goal of limiting GHG emissions to no more than 2 °C by 2050.

Before discussing verification and validation in detail, it is important to point out how these activities came to be defined and practiced. We also distinguish verification and validation from other types of environmental auditing, such as certification of organizations to management systems standards.

As carbon markets developed with the introduction of the Clean Development Mechanism (CDM) and other voluntary and compliance projects, investors required assurance that project emission reductions and removal enhancements were real, additional, permanent, quantifiable, verifiable, and enforceable. The financial accounting profession offered a framework for issuing opinions called ISAE 3000: Assurance engagements other than audits and reviews of historical financial information. The title is a bit complicated. It defines its scope of application by stating what the framework does not address. The title says the framework does not apply to financial information, like annual or quarterly accounts of profit and loss. ("Reviews" are a less intensive activity than an "audit.") Any other topic is fair game, under the right conditions. For our purposes, the key term in the title is "assurance engagements."

Assurance is provided in the form of an "opinion" after the performance of audit techniques. Assurance provides the "intended users" of the opinion with confidence that the verified or validated statement was "fairly stated," and in conformity with "criteria." The words that financial auditors use are chosen very carefully, both for technical reasons and to limit liability for the auditors and their firms. Whether a company is stating its financial results or the quantity of GHGs that its operations emitted, it is very difficult to achieve accuracy to the exact dollar or metric ton. Thus, a statement of GHG emissions may be considered "fair" if it deviates from the exact amount by no more than a pre-defined level, often plus or minus five percent. Misstatements that exceed this level of "materiality" lead to either a modified opinion or to an opinion offering no assurance.

In 2002, the International Organization for Standardization (ISO) undertook to write a standard on validation and verification of greenhouse gas statements. The first edition of ISO 14064 Part 3 was published in 2006. It was subsequently revised and the second edition, published in 2019, has replaced it (ISO 2019). Another ISO standard, ISO 14065, was published in 2007 to specify requirements for validation and verification bodies. This standard is widely used around the world as the basis for the accreditation of bodies, particularly third-party bodies, that offer validation and verification services in the marketplace. The third edition of ISO 14065 was published in 2020 (ISO 2020).

5.2.1 Oversight of Validation and Verification Bodies

Validation and verification bodies play an important role. Like financial auditors, they are typically hired by the management of the organizations whose statements of emissions—or emission reductions and removal enhancements—they validate or verify. Due to this commercial relationship, market participants expect a high level of integrity by firms providing services. Confidence in validation and verification bodies is increased by oversight bodies who check whether validation and verification bodies have applied strict ethical procedures to avoid conflicts of interest. They also assess whether validation and verification bodies have fulfilled their roles impartially, using an evidence-based approach, and fairly presented the results of their activities.

Oversight bodies may be either government authorities or national accreditation bodies. Many countries around the world have national accreditation bodies whose scope of activity includes assessing local validation and verification bodies to the requirements of ISO 14065 and ISO 14064 Part 3.

For an additional level of scrutiny, national accreditation bodies have created a membership organization called the International Accreditation Forum (IAF 2021). This organization establishes multilateral agreements (MLA) among its members whose purpose is to share best practices and conduct peer assessments of the MLA signatories. The goal is to ensure consistent application of accreditation rules and technical standards everywhere in the world.

Validation and verification bodies exist to provide assurance to intended users that the statements made by a "responsible party" are fairly stated. The auditing techniques used to achieve this objective for GHG statements are like those employed by financial auditors who provide opinions on the fair presentation of financial results. What differs is the "subject matter" (GHG emissions, removals, and storage versus statements of financial capital, assets, and liabilities). Validators and verifiers need to be knowledgeable about the subject matter to provide competent professional services. Recognizing this, ISO in 2011 published ISO 14066, Competence requirements for greenhouse gas validation teams and verification teams (ISO 2011).

While validation and verification may be considered a type of "environmental auditing," there is a big difference between the services provided by validators and verifiers and those provided by auditors of management systems. GHG validators and verifiers employ a variety of techniques not commonly used by management systems auditors. Auditors who assess management systems against the requirements of ISO 14001 will need to develop additional skills to undertake assurance engagements in conformity with ISO 14064 Part 3. While management systems auditors focus their inquiry on "conformity with requirements," GHG validators and verifiers use an expanded set of techniques to determine "fair presentation" of data and information. Assurance is obtained only after evidence has been gathered that demonstrates that GHG quantification procedures produced accurate, reliable, and reproducible results.

5.2.2 Types of Engagements

The second edition of ISO 14064 Part 3 specifies requirements for the verification of GHG statements of historical information. ISO 14064 Part 3 defines "GHG statement" as a "factual and objective declaration" that provides the subject matter for verification or validation (ISO 2019, 3.4.3). Examples include organizational inventories that compile information about emissions at the organizational level and mitigation project statements that report on the achievement of emission reductions or removal enhancements that have occurred during a reporting period. When users of this information are interested to know that statements are accurate and can be relied upon, they engage verifiers to provide assurance.

Assuming sufficient records exist and that any discovered material misstatements are corrected, a verifier should be able to provide an opinion at a "level of assurance" that stated information has been presented fairly, where "level of assurance" is defined as a "degree of confidence in the GHG statement" (ISO 2019, 3.6.5). The term "reasonable assurance" means there is a high, but not absolute, degree of confidence that the stated GHG emissions or removals, or emission reductions or removal enhancements, were fairly presented in the organization's, project's, or product's GHG statement. This level of assurance satisfies the objectives of most intended users of verified GHG statements who want to know whether they can rely on the accuracy of reported information. A "limited level of assurance" also may be provided, and this option is explained later in this chapter.

ISO 14064 Part 3 also defines requirements for the validation of GHG statements. Validation is defined in the standard as a "process to evaluate the reasonableness of the assumptions, limitations, and methods that support a statement about the outcome of future activities" (ISO 2019, 3.6.3). The key difference here is that, unlike verification, whose object is historical information, validation evaluates statements related to future activities and their outcomes. Since future outcomes are to some extent uncertain or even unknowable, validation differs significantly from verification. Validators try hard to understand how plans will be realized and what the effect will be of the application of technology, human inputs, and engineered controls. At their best, however, validators can only provide opinions on the reasonableness of the underlying assumptions, limitations, and methods that a project developer has employed to estimate future emissions. The validator notes, however, that plans, and forecasts are not likely to unfold exactly as predicted, and that the differences may be material.

The second edition of ISO 14064 Part 3 takes into account that further subtleties exist beyond the difference between verification and validation. Consider, for example, an organizational inventory that reports "indirect emissions" or a GHG statement about the carbon footprint of a product. In both cases, a responsible party may include information in their statements that lies outside the control of the persons preparing a GHG emissions inventory.

Indirect emissions include imported energy, typically in the form of electricity or steam, and other indirect emissions, which include upstream and downstream supply chain emissions, plus employee emissions attributable to the purpose of the reporting organization (e.g., commuting to work). Indirect emissions may not be verifiable as historical information controlled by an organization, a project developer, or the producer of a good or service. The organization can report kilograms of steam consumed, but if it purchases the steam from a third party, it may not have evidence to support verification of the emissions responsible for making and delivering that steam. In this case, the verifier must determine whether "sufficient and appropriate" information exists to express an opinion on the fair presentation of the emissions.

Indirect emissions associated with the purchase of electricity are common. Organizations typically have access to reliable "activity data" (e.g., kWh consumed) and reliable emissions factors used to quantify such emissions. In such cases, the verifier may conclude that the evidence available is "sufficient and appropriate" and express an opinion on these indirect emissions. Instances arise, however, where historical information may not be available to reach either reasonable or limited assurance on the emissions quantified or the appropriate emissions factor, or both.

Consider, for example, emissions attributed to employee commuting from home to work. Information about employee commuting habits may be gleaned from the results of a questionnaire circulated among employees, but records of the number of trips in various modes of transportation with accurate distances and vehicle types are not available. Moreover, a survey conducted prior to the pandemic of 2020 would likely not accurately represent post-pandemic commuting patterns. A responsible party may deal with situations like this by updating the estimation model periodically and when circumstances change. Information developed by surveys or by other means to develop estimates may still not provide a verifier with evidence that is "sufficient and appropriate" for offering a verification opinion.

ISO 14064 Part 3 provides a new tool called "agreed-upon procedures" (AUP). In circumstances where verifiers do not have access to "sufficient and appropriate" historical information, they choose not to express an opinion providing assurance on that portion of the responsible party's statements. Instead, they use AUP to evaluate specific statements made by responsible parties and to report factual findings determined in accordance with the procedure followed. In an agreed-upon procedures engagement, it is left to the intended users of the procedures to draw their own conclusions from findings presented in the verifier's report. To summarize, the second edition of ISO 14064 Part 3 recognizes that some indirect emissions may not be able to be verified to standards of reasonable or limited assurance and provides an alternate method for dealing with these.

5.2.3 Verification Planning

Verification planning is essential for the success of a verification or validation engagement. This involves understanding objectives, criteria, and scope of the work, choosing the right type of engagement, undertaking a strategic analysis and risk assessment, and determining materiality thresholds. It is completed through document review and (optionally) one or more site visits. Outputs of the planning process include a verification plan and an evidence-gathering plan.

Clause 6 of ISO 14064 Part 3 describes planning requirements that apply to the verification of three types of GHG statements: those for organizational inventories, projects, and products. Planning begins with an understanding of the objectives for the engagement. A verification body may be contracted by a responsible party or client who wishes to obtain assurance that a GHG statement is fairly stated. Assuming the body has access to verifiers with the required competence, the discussion with the client[1] proceeds to a mutual understanding of the engagement scope, criteria, and type of engagement.

Scope defines boundaries for the engagement. These may be organizational in the case of an inventory, geographical or technological in the case of a project, or the system boundaries of a product carbon footprint. Included within boundaries are physical infrastructure, such as facilities, plant, and equipment; activities that transform inputs into outputs; and specific technologies and processes. These physical attributes include sources, sinks, and reservoirs of greenhouse gases of various types, all of which need to be identified. Temporal boundaries (time periods) are also important for verification of historical information and are identified in the scope of the engagement.

---

[1] The term "client" is used to represent the party for whom the verification engagement is performed. The client may also be the responsible party but is not necessarily so. The client may also be a GHG program administrator or stakeholder (ISO 2019).

Having understood the client's objective and scope, the verifier selects the appropriate engagement type. If the GHG statement encompasses only historical information, such as direct emissions from on-site plant and equipment, the verification engagement type suffices. If, however, the scope includes indirect emissions where evidence supporting quantification is lacking, the verifier may suggest to the client that a mixed engagement of verification and agreed-upon procedures is indicated. In the case of a carbon product footprint, the verifier may also realize that the scope includes projected emissions, such as those that will be realized during the product's use and end-of-life stages. An element of this type within a GHG statement calls for the use of validation procedures. Indeed, most carbon footprint statements typically include historical information suitable for verification (gate-to-gate emissions at a manufacturing facility, for example), indirect emissions from suppliers, and projected emissions information related to the activities of users and end-of-life processes. In this case, a mixed engagement may consist of elements of verification, validation, and agreed-upon procedures.

5.2.4 Criteria and Materiality

"Criteria" are defined as the "policy, procedure or requirement used as a reference against which the GHG statement is compared" (ISO 2019, 3.6.10). A note to this definition points out that "criteria may be established by governments, GHG programs, voluntary reporting initiatives, standards, good practice guidance, or internal procedures." For example, an organization reporting on its GHG emissions and removals may develop an inventory based upon the requirements of ISO 14064 Part 1, Specification with guidance at the organization level for quantification and reporting of greenhouse gas emissions and removals (ISO 2018). In this case, the verifier's job consists of verifying both the accuracy of the quantified emissions and the conformity of the responsible party's reporting to the requirements of the standard.

ISO 14064 Part 3 requires the verifier to assess the suitability of the criteria proposed by the client (ISO 2019, 5.1.5). The verifier needs to consider how the client determined the scope of the inventory and its boundaries; the types of GHGs involved and their relevant sources, sinks and reservoirs; quantification methods; and requirements for disclosure. The verifier will find the criteria suitable when they are relevant, complete, reliable, and understandable. Prior to continuing verification activities, the verifier ensures that the chosen criteria are agreed with the client. At the same time, it is made clear that the principles, underlying standards, and GHG program requirements will be applied during the verification, as applicable.

The verifier identifies one or more materiality thresholds for the engagement (ISO 2019, 5.1.7). "Materiality" is the concept that an individual misstatement, or aggregation of misstatements, might influence the decisions of an intended user. Materiality is thought of first in a quantitative sense, as the percentage of deviation of a GHG statement from an accurate quantification of GHG emissions or removals. An overall threshold of materiality may be defined in criteria. In many cases intended users of verification opinions expect the responsible party's GHG statement to be accurate within $\pm 5\%$. Moreover, ISO 14064 Part 3 requires verifiers to set a "performance materiality" threshold for different elements of the GHG statement which may be lower than the overall threshold of materiality defined in criteria. This is done to ensure that, by the end of the verification, the verifier can conclude that the entire GHG statement is accurate within the prescribed threshold (ISO 2019, 6.1.2.1). The concept of materiality is also applied to qualitative aspects of a GHG statement.

Sources for requirements are applicable criteria, such as standards, GHG program requirements, or legal requirements. Thus, if a standard such as ISO 14064-1 requires CO_2 emissions from the combustion of biomass to be reported separately (as it does), and the organization has consolidated these emissions with non-biomass emissions of CO_2, the GHG statement's presentation of quantified CO_2 emissions may be accurate on a reported CO_2 emissions basis, but not conform to the criteria. Generally, verifiers treat nonconformities to criteria as issues to be fixed by the responsible party prior to issuing a verification opinion. However, the verifier may use professional judgment in determining whether a qualitative nonconformity is "material" and not require the responsible party to correct the issue. Depending upon their severity, uncorrected qualitative nonconformities may result in the verifier issuing either an adverse opinion or one that is modified. We address the drafting of opinions later in this chapter.

5.2.5   Strategic Analysis and Risk Assessment

Central to verification planning is the verifier's strategic analysis and risk assessment. At the conclusion of these two activities, the verifier will issue a verification plan and an evidence-gathering plan that the verification team will execute during the remainder of the verification engagement.

Strategic analysis provides the verifier with an understanding of the activities and complexity of the organization, project, or product that is the subject of the responsible party's GHG statement. The requirements for the scope of the strategic analysis are provided in an itemized list (ISO 2019, 6.1.1) that includes:

– Relevant sector information
– The nature of the operations
– Applicable regulatory requirements
– The scope of the GHG statement and related boundaries
– Sources, sinks, and reservoirs applicable to the organization, project, or product
– The GHG statement and prior period reporting information
– Data management and data control systems and controls.

This is a partial list. ISO 14064 Part 3 includes at least 20 different items that verifiers shall consider during the strategic analysis step.

5.2.6 Risk Assessment

The results of the strategic analysis inform the verifier's assessment of audit risk. Audit risk is the risk that the verifier may inappropriately issue an unmodified or modified opinion in cases where the verifier failed to detect material misstatement due either to error or fraud. Although strategic analysis and risk assessment (ISO 2019, 6.1.2) are presented sequentially in ISO 14064 Part 3, they are interrelated and iterative. With that in mind, the standard asks verifiers of emissions and removals to assess three types of risk (ISO 2019, 6.1.2.2):

- Inherent risks
- Control risks
- Detection risks.

The verifier then assesses the relevance of each type of risk to the following characteristics of emissions or removals: occurrence, completeness, accuracy, cut-off, and classification (ISO 2019, 6.1.2.2).

Occurrence is the risk that reported emissions or removals were not "real." Completeness defines another type of risk. Determining the accuracy of stated GHG emissions and removals is a primary objective of verification. Cut-off is the risk that emissions or removals will not be allocated to the proper time period. Classification is the risk that an emission will not be properly categorized.

Understanding the likely causes for these risks allows the verifier to categorize them as "inherent" or "control" risks. Inherent risks arise from the nature of the processes, technologies, or operations that can result in GHG emissions or removals. Control risks, on the other hand, arise from an organization's attempt to manage inherent risks through the design and implementation of procedural, engineering, or software controls. The higher the risk, the more evidence the verifier should gather to determine whether emissions or removals are accurately reported. In designing evidence-gathering activities, the verifier also considers "detection" risk, which is the risk that the verifier will not find evidence of misstatement where misstatement exists. The results of the verifier's assessment of risks are reflected in the verification plan and the evidence-gathering plan.

As applicable, the risk assessment shall consider the following "considerations" (ISO 2019, 6.1.2.3)[2]:

- Likelihood of intentional misstatement in the GHG statement
- The relative effect of emission sources on the overall GHG statement and materiality
- Likelihood of omission of a potentially significant emission source
- Whether there are any significant emissions that are outside the normal course of business for the responsible party or that otherwise appear to be unusual

---

[2] Clause 6.1.2.3 ©ISO. This material is reproduced from ISO 14064–3:2019, with permission of the American National Standards Institute (ANSI) on behalf of the International Organization for Standardization. All rights reserved.

- Nature of operations specific to an organization, facility, project, or product
- Degree of complexity in determining the organizational or project boundary or product system boundary and whether related parties are involved
- Changes from prior periods
- Likelihood of non-compliance with applicable laws and regulations that can have a direct effect on the content of the GHG statement
- Significant economic or regulatory changes that might impact emissions and emissions reporting
- Selection, quality, and sources of GHG data
- Level of detail of the available documentation
- Nature and complexity of quantification methods
- Degree of subjectivity in the quantification of emissions
- Significant estimates and the data on which they are based
- Characteristics of the data management information system and controls
- Apparent effectiveness of the responsible party's control system in identifying and preventing errors or omissions
- Controls used to monitor and report of GHG data
- The experience, skills, and training of personnel.

5.2.7   Site Visits

The second edition of ISO 14064 Part 3 adds explicit requirements for site visits by the verification team and describes in detail what types of activities must be performed during these visits.

Verifiers shall conduct site visits (ISO 2019, 6.1.4.2)[3] in all the following circumstances:

- An initial verification
- A subsequent verification for which the verifier does not have knowledge of the prior verification activities and results
- A verification where there has been a change of ownership of a site or facility and the emissions, removals, and storage of the site or facility are material to the GHG statement
- Misstatements are identified during the verification that indicate a need to visit a site or facility
- There are unexplained material changes in emissions, removals, and storage since the previous verified GHG statement.
- The addition of a site or facility of GHG sources, sinks, and reservoirs that are material to the GHG statement
- Material changes in scope or boundary of reporting

---

[3] Clause 6.1.4.2 ©ISO. This material is reproduced from ISO 14064–3:2019, with permission of the American National Standards Institute (ANSI) on behalf of the International Organization for Standardization. All rights reserved.

– Significant changes in the data management involving the specific site or facility.

The verifier employs a risk-based approach to identify the need to visit facilities and to select the number and location of facilities to visit (ISO 2019, 6.1.4.1).[4] This process takes into consideration the following:

– Results of the risk assessment and efficiencies in collecting evidence
– Number and size of sites and facilities associated with the organization, project, or product
– Diversity of activities at each site and facility contributing to the GHG statement
– Nature and magnitude of the emissions at different sites and facilities, and their contribution to the GHG statement
– Complexity of quantifying emissions sources generated at each relevant site or facility
– Degree of confidence in the GHG data management system
– Risks identified through the risk assessment indicating the need to visit specific locations
– Results of prior verifications or validations, if any.

Verifiers visit facilities to assess a variety of operations, activities, and systems (ISO 2019, 6.1.4.3). Observation assists the verifier to identify GHG sources, sinks and reservoirs, and physical infrastructure and equipment. Observation also helps the verifier understand process flows and data management, including control systems. Key measurement devices can be assessed for correct installation, operation, and data recording. The scope of GHG reporting and boundaries of the facility can be checked against information provided in the GHG statement. Monitoring and measurement activities can be observed, and personnel with key data collection responsibilities interviewed. The verifier also uses the facility visit to assess the design and implementation of quality control/quality assurance procedures and to consult records related to calculations and assumptions made in developing GHG data.

Knowledge of the responsible party's operations, its control environment, and its data collection methods informs the verification plan (ISO 2019, 6.1.5). This document is one of two key outputs of the strategic analysis and the risk assessment. It provides a schedule of verification activities that will take place, either off-site or at facilities, during the verification engagement.

The second key output of the strategic analysis and risk assessment is the evidence-gathering plan (ISO 2019, 6.1.6). This document defines in greater detail the type and extent of evidence-gathering activities that members of the verification team will execute during their verification engagement. The evidence-gathering plan is developed to ensure that the verification team determines whether the GHG statement conforms to criteria, including by considering the principles of the standards or GHG program requirements which apply to the GHG statement.

---

[4] Clause 6.1.4.1 ©ISO. This material is reproduced from ISO 14064–3:2019, with permission of the American National Standards Institute (ANSI) on behalf of the International Organization for Standardization. All rights reserved.

Planning, in other words, is a rigorous process. ISO 14064 Part 3 reflects this by devoting nine pages of the standard to this phase of the verification activity. By contrast, the "execution" phase is presented in only two sentences.

5.2.8 Transitioning from Planning to Execution

Planning a verification engagement anticipates the execution phase. During planning, the verifier identifies criteria, becomes familiar with the GHG information system and controls, and understands the responsible party's data aggregation process. With this information at hand, the verifier can identify appropriate verification activities which are detailed in the evidence-gathering plan. According to ISO 14064 Part 3 (ISO 2019, 5.3),[5] these include:

- Observation
- Inquiry
- Analytical testing
- Confirmation
- Recalculation
- Examination.

Each of these activities is available to the verifier whose objective is to reduce the risk of material misstatement to an acceptably low level. Thus, observation may be used to assess the risk that operations and activities performed at a facility include GHG sources that have not previously been identified. Inquiry may be used to obtain information about controls, or to gather evidence about regulatory compliance. Analytical testing typically is used to determine the accuracy of data, such as the total amount of fuel purchased and consumed.

Confirmation occurs when the verifier contacts an external party, such as a supplier, to confirm the accuracy or completeness of data compiled from records. Confirmation may also involve contacting a regulatory body to confirm the compliance status of a responsible party. Recalculation confirms the accuracy of the responsible party's statement of emissions, emission reductions, or removals, as applicable. And examination involves looking at specific items, such as flow meter elements or data transmitters, to determine their maintenance or calibration status.

In the execution phase of the engagement, the verifier executes analytical tests of the effectiveness of the GHG information system and controls. The verifier is then ready to perform tests of the accuracy and completeness of the GHG data. In the planning phase, the verifier designs tests of both the information system and controls, and of data. In the execution phase, the verifier tests selected data and information to confirm occurrence, accuracy, completeness, cut-off, and classification.

---

[5] Clause 5.3 ©ISO. This material is reproduced from ISO 14064–3:2019, with permission of the American National Standards Institute (ANSI) on behalf of the International Organization for Standardization. All rights reserved.

Ideally, the verifier matches his assessment of risks with analytical activities designed to determine whether, after testing, any residual risk of material misstatement has been reduced to an acceptably low level. But the "performing" of the analytical procedures is an element of the "execution" phase of the engagement, which demonstrates how intricately linked and iterative the planning and execution phases of the engagement are.

ISO 14064 Part 3 (ISO 2019, 5.3 g–m) identifies eight different verification techniques that can be included in the designed procedures of an evidence-gathering plan:

- Tracing
- Retracing
- Control testing
- Sampling
- Facility visits
- Estimate checking
- Cross-checking
- Reconciliation.

Tracing is a test applied to data in the GHG statement. For example, direct emissions from reported fuel consumption are compared to the volumes of fuel supported by purchase records. This test tends to identify potential overstatements in GHG reporting. Retracing starts from the source data and tracks that information forward to inclusion in the GHG statement. It tends to identify understatements of emissions.

Control testing focuses on determining the effectiveness of the reporting entity's strategies for ensuring accurate and complete reporting. Types of controls include administrative, engineering, and software. The administrative category includes training, supervision, internal audit, and management review. Engineering controls include labeling, maintenance and calibration of measurement devices, perimeter security and access control, and alarm systems to identify potential malfunction of systems. Controls typically designed into software applications may include input, transformation, and error checking routines; checks on the transfer of information between different systems; and output controls surrounding the distribution of GHG information and comparisons between input and output information.

Sampling is an important verification technique since it is usually not feasible for a verifier to examine 100% of the available data. Data of a particular type, such as the amount of HFCs used in an organization's refrigeration equipment, form a "population" of data. To sample means examining a subset of that population of data to check the accuracy and completeness of reported fugitive emissions. One type of sampling is random, in which any item in the population has an equal chance of being chosen. Other types include strategic, where a sample is specifically selected because of indications that the data may present a high risk of misstatement; systematic, where the sample has a specified sampling interval; stratified, where the population has been subdivided into subpopulations; convenience, where the sample is readily available to the verifier; and unstructured, where the sample has no structural technique. Verifiers should avoid convenience sampling and unstructured sampling, as these types present the lowest amount of methodological rigor and sampling risk—the risk that the results from sampling may return either false positives or false negatives—is relatively high.

Estimate checking should be performed by verifiers when elements of the GHG statement that are material have been developed using estimates rather than primary data. For example, a manufacturer who reports the indirect transportation emissions from raw materials suppliers may estimate the distance traveled by supply trucks or rail cars. When assessing the accuracy and completeness of the estimate, the verifier should examine the basis for the estimated distances and consider other information such as whether the truck or rail car returns empty or loaded.

Cross-checking is a technique that compares the responsible party's basis for calculation of GHG emissions, emission reductions, or removals with an alternate method. Cross-checking is important to ensure that verifiers do not confirm, through recalculation, systemic errors that a responsible party may have made. Such errors may be due to many reasons, such as omission of sources, improper conversion of units, programming errors in database queries, or the application of inappropriate emission factors, just to name a few. Cross-checking, when done at the aggregated level of reporting, such as by facility, can reveal patterns that typically persist year after year. For example, Plant A is known to operate about 5% more efficiently than Plant B. If unexplained changes occur in these relationships during a given reporting year, the verifier should investigate the causes.

Finally, reconciliation verifies that the line-by-line accounting of a GHG inventory, project, or product adds up to the subtotals and totals in a GHG statement. For example, in GHG projects, the ultimate equation typically is stated as "$ER = BE - PE$," where ER = emission reductions, BE = baseline emissions, and PE = project emissions. Reconciliation in this context involves checking that BE minus PE equals ER. As part of the process, the verifier should check that the aggregated totals for these terms accurately sum the amounts reported at facility or business unit levels. The amounts are "reconciled" when the subtotals of both BE and PE are added up, confirmed, and compared to ER.

Verifiers typically will devote more time on the assessment of GHG information systems and controls. This is necessary as the greater the amount of GHG data and information to be verified, the lower the percentage of information the verifier can check. Verifiers need to test some minimum percentage of information in every engagement, but the percentage of information tested can vary from single digits where the population of available information is very large to twenty, forty, or even larger percentages where the population of available information is relatively small. Where only small percentages of the total GHG data available to check is tested, the verifier must ensure that the risk of misstatement from errors in the GHG data and information system is very low and that other controls established by the organization are properly designed, implemented, and effective.

5.2.9 Execution and Completion of Verification Activities

Clause 6.2 of ISO 14064 Part 3 is short, as the planning clause describes in rather full detail the key elements of the verification process: strategic analysis and risk assessment, design and execution of verification procedures, and approval of initial plans and changes to the verification strategy. This clause states that the verifier shall execute the verification in accordance with the verification plan and evidence-gathering plan. The verifier shall also assess material changes that the responsible party makes to the GHG statement (ISO 2019).

Clause 6.3 completes the verification section of ISO 14064 Part 3. Its requirements address the final steps of evaluation, communication, and analysis that the verifier must complete prior to turning over the verification engagement documentation for independent review and issuance of the verification opinion. At this stage of the engagement, the verifier evaluates where there have been any changes in risks and materiality that may have occurred over the course of the verification. If yes, the verifier must evaluate whether any high-level analytical procedures that have been applied remain representative and appropriate.

The verifier determines whether the evidence collected is sufficient and appropriate to reach a conclusion. If the verifier determines there is insufficient or inappropriate evidence, the verifier must develop additional evidence-gathering activities. Misstatements also are evaluated to determine whether they remain below the threshold of materiality and to document them.

In the completion phase of the engagement, the verifier ensures that documentation of the verification process is complete and understandable. In preparation for the opinion drafting phase, the verifier evaluates and documents errors and omissions that singly or in aggregate constitute material misstatements or nonconformities with criteria. These steps accomplished, the verifier communicates material misstatements to the responsible party. The verifier may also communicate immaterial misstatements to the responsible party. Based on all the evidence collected, evaluated, and documented, the verifier reaches conclusions from the engagement and drafts a verification opinion, considering any misstatement or nonconformity left uncorrected by the responsible party.

5.2.10 Drafting an Opinion

A verification opinion expresses the conclusions of a verifier with respect to the accuracy of historical information included in the responsible party's statement. It provides an intended user with assurance concerning the fair presentation of the responsible party's GHG statement and the conformity of that statement to criteria. The phrase "fair presentation" is used to avoid misrepresenting the precision with which the verifier expresses an opinion on accuracy. A "verified" number may not represent exactly a responsible party's emissions, but it is accurate within a defined threshold of materiality and thus may vary from the "true" number by some percentage. This variance, by definition, is deemed to be sufficiently small as to not affect any decisions that the intended user of the verification opinion may make when relying upon the statement.

The verifier is responsible for drafting an opinion addressed to intended users. According to clause 9.2 (ISO 2019), opinions consist of one of the following types:

- Unmodified (the verifier expresses no reservations about the opinion or the quality of the evidence that supports it)
- Modified (the verifier brings to the attention of the intended user that some misstatements exist in the GHG statement, but that these are not significant enough to warrant an adverse opinion or
- Adverse (the verifier has determined that the evidence collected and evaluated during the verification does not support the expression of an unmodified or modified opinion).

A disclaimer of opinion is not an opinion. Rather, it is a written statement addressed to the client explaining that the verifier was not able to reach a conclusion and express an opinion. It provides formal notice to the client of the cessation of the verification engagement.

Details of the work performed are included in a verification report.

5.2.11   Limited Assurance

The description of verification activities in clause 6 of ISO 14064 Part 3 (ISO 2019) assumes that the verifier and client have agreed on an engagement at the reasonable level of assurance. Some verifications are performed at a *limited* level of assurance, which provides intended users of the opinion a lower degree of confidence in the verified results. The requirements for a limited level of assurance engagement are described in Annex A of Part 3. In this engagement type, verifiers do not design and apply as many evidence-gathering activities or pursue evidence trails to the same depth as is done in a reasonable assurance engagement.

Limited level of assurance engagements is commonly performed during interim reporting periods. Once a verifier has performed verification at the reasonable level of assurance and understood and tested for effectiveness the responsible party's information system and controls, subsequent verifications may be performed at the limited level of assurance. GHG programs may specify the number of years at which limited level of assurance engagements may be performed. Some may require verifiers to perform a reasonable level of assurance engagement every three years, for example. It is also common for GHG programs to list exceptions to the general rule and mandate reasonable level of assurance engagements under the following scenarios:

- A change in verification bodies
- The issuance of an adverse opinion in the previous reporting period, or of a modified opinion
- A change of operational control of the reporting entity in the previous year
- Significant changes in GHG sources or emissions (e.g. greater than 25%).

Limited level of assurance verification engagements follow the same general process as reasonable level of assurance engagements, with the following exceptions. Strategic analysis does not need to include a detailed assessment of the design, existence, and effectiveness of controls because there is an underlying assumption that the controls are reliable. Risk assessments are performed on the GHG statement as a whole and are not as detailed as those performed for reasonable levels of assurance. This means that the verifier concentrates on the "top line" statements of emissions and does not need to identify risks at a more detailed level.

The verifier designs evidence-gathering activities that address all material items in the GHG statement and focuses on areas where material misstatements are likely to arise. The activities employed by the verifier consist primarily of inquiry and analytical procedures to obtain sufficient and appropriate evidence. If the verifier becomes aware of potential material misstatements, the verifier designs appropriate evidence-gathering activities to investigate these further. These activities must be conclusive enough to allow the verifier to reach a conclusion about the potential material misstatements.

Unless a verifier has prior knowledge of the facility or site's data management process and controls, he or she should visit the facility or site that aggregates data for the GHG statement. Site visits are most commonly made to the corporate headquarters of a reporting organization, where individuals are available for face-to-face interviews and records are readily examined.

The opinion includes a statement that the verification activities applied in a limited level of assurance engagement were less extensive in nature, timing, and extent than in a reasonable level of assurance engagement. Annex A.5 (ISO 2019) stipulates that opinions are expressed using the negative form, for example:

"Based on the process and procedures conducted, there is no evidence that the GHG statement.

- Is not materially correct and is not a fair presentation of GHG data and information, and
- Has not been prepared in accordance with related International Standards on GHG quantification, monitoring, and reporting, or to relevant national standards or practices."

5.3   Validation

Until now, this chapter has focused on verification. Verification always involves historical information, and the objective of the verifier is to reach conclusions based on sufficient and appropriate evidence. This permits the verifier to express an opinion with reasonable or limited assurance on the "fair presentation"—the accuracy—of the responsible party's GHG statement. The subject matter of validation, on the other hand, is estimates or forecasts of data and information that will occur in the future. This includes project validation, where the objective of the validator is to reach a conclusion on the reasonableness of the assumptions, limitations, and methods that were used to predict whether a mitigation project would deliver anticipated emission reductions or removal enhancements. It also includes estimates about the use phase or the end-of-life phase of a carbon footprint.

ISO 14064 Part 3 dedicates clause 7 to the process of validation (ISO 2019). Clause 7 describes a set of activities that should be performed for validating mitigation projects and reaching conclusions. The clause begins, as does clause 6, with a planning phase that starts with strategic analysis. A validator is required to have a sufficient understanding of the GHG-related activity and its relevant sector information to plan and conduct the validation. This enables the validator to:

- Identify the types of potential material misstatements and their likelihood of occurrence
- Select the evidence-gathering procedures that will provide the validator with a basis for his or her assessment and conclusions.

In accordance with clause 7.1.1,[6] the strategic analysis considers, as applicable:

(a)   Relevant sector information
(b)   The nature of operations
(c)   The requirements of the criteria, including applicable regulatory and/or GHG program requirements
(d)   The intended user's materiality, including the qualitative and quantitative components
(e)   The likely accuracy and completeness of the GHG statement
(f)   The proper disclosure of the GHG statement
(g)   The scope of the GHG statement and related boundaries
(h)   The time boundary for data
(i)   Emissions sources, sinks, and changing reservoirs and their contribution to the overall GHG statement
(j)   Appropriateness of quantification and reporting methods, and any changes
(k)   Sources of GHG information
(l)   Data management information system and controls

---

[6] Clause 7.1.1 ©ISO. This material is reproduced from ISO 14064–3:2019, with permission of the American National Standards Institute (ANSI) on behalf of the International Organization for Standardization. All rights reserved.

(m)   Management oversight of the responsible party's reporting data and supporting
      processes
(n)   The availability of evidence for the responsible party's GHG information and
      statement
(o)   The results of sensitivity or uncertainty analysis
(p)   Other relevant information.

As with verification, the validator identifies quantitative materiality thresholds
with respect to elements of the GHG statement. Then the validator evaluates whether
the assumptions provided in or underlying the GHG statement comply with the
criteria, and whether the projected emissions or removals are appropriate. Key factors
here are the applicability of assumptions and the quality of data used in making the
estimate. From this review, the validator develops evidence-gathering procedures
that test the effectiveness of the methods used in formulating the estimate. And
the validator develops his or her own estimate as a check on the reliability of the
responsible party's estimation method.

Clause 7.1.4 in ISO 14064 Part 3 (ISO 2019) is titled "assessment of GHG-related
activity characteristics." The thirteen characteristics of projected information in this
clause clearly set apart validation from verification.

The first characteristic is recognition. This addresses the question of whether the
intended user of the validation opinion recognizes the GHG-related activity. In mitigation projects, this means meeting any eligibility requirements, including geographical and temporal restrictions, and determining that the GHG-related activity is real,
quantifiable, verifiable, permanent, and enforceable.

Ownership addresses whether the responsible party owns or has the right to claim
emission reductions or removal enhancements expressed in the GHG statement.

Validators need to assess whether GHG boundaries associated with the GHG
statement are appropriate and whether all relevant sources, sinks, and reservoirs
within them have been identified and considered.

Baseline selection is a key characteristic for mitigation projects. This step in
validating a project goes to the core issue of whether a project will be recognized
by the intended user of the validation opinion. The validator looks for a baseline
determination process that is credible, documented, and repeatable, and one that is
appropriate for the GHG-related activity. The methods used for baseline selection
should consider conservativeness, uncertainty, common practice, and the operating
environment of the GHG-related activity.

Activity measurements should be appropriate for the operational conditions and
the associated activity levels used in the GHG quantification methodology. Once
the project is validated, these measurements will be used to determine emission
reductions or removal enhancements, and they must be capable of producing data
that are accurate, complete, and conservative.

"Secondary effects" is a new term in ISO 14064 Part 3. It replaces the term "leakage" that was adopted by the executive board of the Clean Development Mechanism
and has been widely used since. For GHG-related activities that assert emission
reductions or removal enhancements, validators assess the GHG-related activity to

determine whether material economic effects during the GHG statement period will change emissions outside the GHG-related activity boundary. If the GHG-related activity has accounted for secondary effects, the validator assesses the project's adjustments for completeness and accuracy.

The validator assesses whether the selected quantification methodologies and measurements (monitoring) are acceptable to the intended user. Monitoring needs to be accurate and reliable, conservative, appropriately applied, and disclosed, particularly when operational ranges, operational conditions, or assumptions have not been met.

The validator assesses the GHG information management system and controls to determine the extent to which they can be relied upon in future verification. Key elements include whether measured and monitored data correspond with GHG calculations, and whether such inputs as emissions factors, unit conversions, and global warming potentials are appropriate and meet criteria. Other factors related to this characteristic include the appropriateness of record keeping and reporting practices, data collection and control operation frequencies, and data backup and retrieval routines. Overall, the validator assesses whether elements of the data management process increase the risk of misreporting and meet the needs of the intended user.

For GHG-related activities that assert emission reductions or removal enhancements, the validator assesses the functional equivalence of the project and baseline. The validator assesses both quantitative and qualitative aspects of functional equivalence and the comparability of the scope of the GHG-related activity boundaries. The validator identifies and documents the functional unit used for the quantitative assessment.

The validator confirms the calculations used in the GHG statement, including emissions factors and global warming potentials, and their consistency with criteria.

If applicable, the validator evaluates the future estimates in the GHG statement. This evaluation includes the proposed approach and assumptions inherent in the projection, the applicability of scope of the projection to the proposed GHG-related activity, and the sources of data and information used in the projection including their appropriateness, completeness, accuracy, and reliability. For GHG-related activities that assert emission reductions or removal enhancements, the validator assesses the comparability between the baseline and the proposed project, including the consistency of assumptions and boundaries across the GHG statement period.

The validator assesses whether the uncertainty in the GHG statement affects disclosure or the ability of the validator to arrive at a conclusion. This includes the identification of uncertainties that are greater than expected and the effect of the identified uncertainties on the GHG statement.

The validator identifies assumptions with high potential for change and assesses whether these changes are material to the GHG statement. This process is called sensitivity analysis.

The remainder of clause 7 is very similar to the requirements of clause 6. The validator develops a validation plan and an evidence-gathering plan, executes these and amends them as needed during the course of the validation engagement. After completion of validation activities, the validator evaluates the GHG statement for material misstatement and conformity with criteria. The GHG statement is also evaluated for proper disclosure of information in accordance with criteria and the needs of the intended user.

As with verification, a validation report is required. The validator drafts an opinion of one of the following types: unmodified, modified, or adverse. A validation opinion may also be disclaimed.

In contrast to verifications, validation opinions do not offer assurance. They are expressed with respect to the "reasonableness" of the basis for the hypothetical (modeled) or forecast information. The word assurance is not used in a validation opinion. Instead, a validator states that, based on his or her validation activities, the assumptions, limitations, and methods used by the responsible party provide a reasonable basis for the forecast information.

5.3.1 Independent Review

Clause 8 of ISO 14064 Part 3 requires an independent review of both verification and validation engagement documentation. The review is conducted by personnel who are competent and different from the persons who conducted the verification or validation. The list of elements of the engagement the independent reviewer shall evaluate include (ISO 2019)[7]:

(a) The appropriateness of team competencies
(b) Whether the validation/verification has been designed appropriately
(c) Whether all validation/verification activities have been completed
(d) Significant decisions made during the validation/verification
(e) Whether sufficient and appropriate evidence was collected to support the opinion.
(f) Whether the evidence collected supports the opinion proposed by the validation/verification team.
(g) GHG statement and the validation/verification opinion and
(h) Whether the verification/validation was performed according to ISO 14064 Part 3, including whether:

- The risk assessment, verification/validation plan, evidence-gathering plan address the objectives, scope, and level of assurance
- For verification:

  - The evidence-gathering activities address the risks identified.
  - A data trail has been established for material emissions, removals, and storage

---

- For validation:

  - The evidence-gathering activities address the GHG-related activity characteristics.

- Verification/validation team decisions are supported be sufficient and appropriate evidence.
- Any restatements have been adequately assessed.
- The GHG statement is in accordance with the criteria.
- Significant issues have been identified, resolved, and documented.

The independent reviewer communicates with the verification/validation team when the need for clarification arises. The verifier/validation team is responsible for addressing any concerns raised by the independent reviewer. The independent reviewer documents the results of the review.

5.3.2   Issuance of Opinion

After completion of the validation/verification, the verifier or validator issues an opinion containing the following elements (ISO 2019, clause 9.3)[8]:

(a)   Identification of the GHG-related activity (e.g., organization, project or product)
(b)   Identification of the GHG statement, including the date and period covered by GHG statement
(c)   Identification of the responsible party and a statement that the GHG statement is the responsibility of the responsible party
(d)   Identification of the criteria used to compile and assess the GHG statement
(e)   A declaration that the verification or validation of the GHG statement was conducted in accordance with ISO 14064 Part 3
(f)   The verifier's conclusion including level of assurance
(g)   The validator's conclusion
(h)   The date of the opinion.

In the case of third-party verifiers or validators, the opinion may contain statements that limit liability for the issuer. Modified, disclaimed, and adverse opinions must present the reasons that led to the selection of the opinion type. If a verifier or validator issues a modified opinion, and the reason for the modification is quantitative misstatement, the verifier or validator must include the value of the quantified misstatement in the opinion and describe its effect on the GHG statement. Where the

---

[8] Clause ©ISO. This material is reproduced from ISO 14064–3:2019, with permission of the American National Standards Institute (ANSI) on behalf of the International Organization for Standardization. All rights reserved.

GHG statement includes a forecast of future emission reductions/removal enhancements, the GHG opinion shall explain that actual results may differ from the forecast as the estimate is based on assumptions that may change in the future.

5.3.3 Facts Discovered After the Verification/Validation

Should facts or new information that could materially affect the verification or validation opinion be discovered after the date the opinion is issued, the verifier or validator communicates the matter as soon as practicable to the responsible party, the client, or the program owner. The verifier or validator may also communicate to other parties the fact that reliance of the original opinion may now be compromised given the new facts or information (ISO 2019, clause 10).

5.4 Agreed-Upon Procedures

A new feature of ISO 14064 Part 3 is Agreed-upon procedures (AUP). Annex C describes how verifiers may use the activities and techniques of verification when intended users do not require an opinion. It is for specialized applications only and is not a substitute for an assurance engagement. Indeed, the ISO working group experts added language in the introduction to Annex C that states that "Verification and validation of GHG statements developed in accordance with ISO 14064 Part 1, ISO 14064 Part 2 and ISO 14067 are performed in accordance with clauses 5 to 10 from this document and AUP shall not be used for this purpose." The intent of the expert group is also illustrated in the standard's Fig. 5.2. The relationship of AUP to verification engagements that provide assurance through the issuance of an opinion, and validation engagements that express an opinion, is clearly distinguished:

AUP are useful in the following circumstances (ISO 2019, C.2)[9]:

(a) GHG programs that specify agreed-upon procedures rather than assurance
(b) Specific indirect emissions and removals (indirect emissions in inventories; upstream and downstream emissions and removals for product life cycles)
(c) Compliance to specifications
(d) Greenhouse gas information and data management and controls.

Let us illustrate how AUP may be used in some of these cases.

---

[9] Clause ©ISO. This material is reproduced from ISO 14064–3:2019, with permission of the American National Standards Institute (ANSI) on behalf of the International Organization for Standardization. All rights reserved.

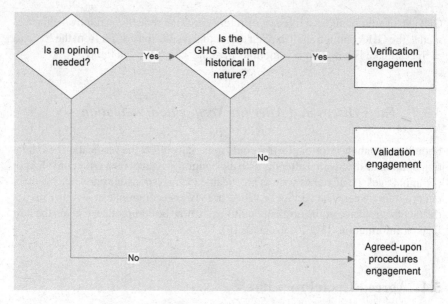

Fig. 5.2 Decision process for engagement type (ISO 2019, Fig. 2) ©ISO. This material is reproduced from ISO 14064-3:2019, with permission of the American National Standards Institute (ANSI) on behalf of the International Organization for Standardization. All rights reserved

5.4.1 Certain Indirect Emissions/Removals

Until the publication of the second edition of ISO 14064 Part 3, verifiers who encountered statements about indirect emissions that were not backed up with "sufficient and appropriate" evidence had only one choice: agree to include them within the scope of verification and lower their standards for evidence, or specifically exclude them from the engagement. Oftentimes indirect emissions from an organization's supply chain or downstream customers were not inventoried at all because the first edition of ISO 14064-1 only required reporting of direct emissions and energy indirect emissions. The reporting of "other indirect emissions" was optional, and these were frequently omitted by responsible parties.

The second edition of ISO 14064-1 charts a new direction, because responsible parties are now expected to report on their "significant" indirect emissions no matter where in the value stream they occur. Going forward, verifiers will face increased instances where statements about upstream and downstream indirect emissions are made. AUP gives verifiers a new tool. It allows them to report factual findings about emissions that could not otherwise be verified because sufficient and appropriate evidence substantiating their occurrence, completeness, accuracy, cut-off, and classification could not be provided by the responsible party.

5.4.2 Compliance to Specifications

Another use for AUP is to provide information to intended users about compliance to specifications. Consider a facility that wishes to destroy ozone depleting substances (ODS) in a GHG mitigation project. In accordance with some methodologies, such facilities must demonstrate that they meet the Code of Good Housekeeping established by the Montreal Protocol's Technology and Economic Assessment Panel (TEAP) in order to qualify as a "destruction facility." The facility asks a verification body with expertise in ODS verification to use AUP to issue factual findings on the facility's compliance with the Code of Good Housekeeping. The verification body is not asked to provide an opinion, but the destruction facility itself and its customers are able to draw conclusions from the reported findings about its compliance with the TEAP requirements.

5.4.3 GHG Information and Data Management and Controls

A responsible party may wish to gather information, for internal purposes, about GHG information and data management and controls. Consider the case of a merger of two companies, where the acquiring company has little knowledge of the GHG information gathered by its new division and the extent and type of data management and controls that it has established. The information is needed for planning purposes, so the responsible party asks its internal verifier to use verification activities and techniques and report on his findings. The results of AUP thus inform planning for the integration of GHG information, data management, and controls by the merged companies.

AUP, as stated above, are not a substitute for a verification engagement that results in an opinion. They are a complementary activity that can be used when an intended user agrees on evidence-gathering activities and takes responsibility for the procedures and for their use. Verifiers do not undertake an agreed-upon procedures engagement when the intended user does not agree to the content and the sufficiency of the procedures (ISO 2019, C.1).

Prior to accepting the AUP engagement, the verifier must determine whether the subject can be measured in a reasonably consistent manner, and whether executing agreed-upon procedures will produce reasonably consistent results. The verifier needs to determine that evidence required for executing the agreed-upon procedures is likely to exist, and that such evidence will form the basis for factual findings. The verifier should not agree to perform procedures that are subjective and open to varying interpretations (ISO 2019, C.3.3).

The verifier executes agreed-upon procedures in the same manner as he performs verification activities and techniques. The nature, extent, and timing of the procedures may be as limited or extensive as the intended user specifies. Appropriate AUP may include (ISO 2019, C.5.1)[10]:

– Execution of a sampling technique after agreeing on relevant parameters
– Inspection of specified documents evidencing certain types of measurements
– Confirmation of specific information with third parties
– Comparison of documents, schedules, or analyses with certain specified attributes
– Performance of specific procedures on work performed by others
– Performance of mathematical computations.

The AUP process is flexible and may be applied to many different types of situations. However, there are inappropriate ways to use AUP. These include (ISO 2019, C.5.2)[11]:

– Simply reading work performed by others that only describes or contains their findings with no other supporting material
– Evaluating the competency or objectivity of another party
– Obtaining an understanding about a particular subject
– Interpreting documents outside the scope of the verifier's professional expertise.

An independent review process may be completed prior to the presentation of AUP results. All results are presented as factual findings with no conclusion or opinion (ISO 2019, C.5.3).

Questions for Readers

1. Why is the use of activity data and emissions factors the dominant method for quantifying GHG emissions?
2. How important is calibration and maintenance of measurement and monitoring devices in ensuring accurate quantification of GHG emissions?
3. In what ways does validation differ from verification?
4. In what ways does a limited assurance verification engagement differ from a reasonable assurance verification?
5. How do agreed-upon procedures differ from verification?

---

[10] Clause ©ISO. This material is reproduced from ISO 14064–3:2019, with permission of the American National Standards Institute (ANSI) on behalf of the International Organization for Standardization. All rights reserved.

[11] See Footnote 10.

References

ECCC (2019) 2019 National inventory report 1990–2017: greenhouse gas sources and sinks in Canada, Part 2. https://unfccc.int/documents/194925. Accessed 25 May 2021

IAF (2021) International accreditation forum. https://www.iaf.nu/. Accessed 10 June 2021

ISO (2011) Greenhouse gases—competence requirements for greenhouse gas validation teams and verification teams. https://www.iso.org/standard/43277.html. Accessed 12 June 2021

ISO (2015) ISO 14001, Environmental management—environmental management systems—requirements with guidance for use. https://www.iso.org/standard/60857.html. Accessed 25 May 2021

ISO (2018) ISO 14064 greenhouse gases—part 1: SPECIFICATION with guidance at the organization level for quantification and reporting of greenhouse gas emissions and removals. https://www.iso.org/standard/66453.html. Accessed 14 June 2021

ISO (2019) ISO 14064 greenhouse gases—part 3: specification with guidance for the verification and validation of greenhouse gas statements. https://www.iso.org/standard/66455.html. Accessed 30 June 2020

ISO (2020) ISO 14065, general principles and requirements for bodies validating and verifying environmental information. https://www.iso.org/standard/74257.html. Accessed 12 June 2021

Lütken S et al (2012) Measuring reporting verifying. https://www.transparency-partnership.net/sites/default/files/2012_unep_risoe_mrv_a_primer_on_mrv_for_namas.pdf. Accessed 4 Dec 2020

US EPA (2020a) Inventory of U.S. Greenhouse Gas Emissions and Sinks Fast Facts and Data Highlights. https://www.epa.gov/ghgemissions/inventory-us-greenhouse-gas-emissions-and-sinks-fast-facts-and-data-highlights. Accessed 30 June 2020

US EPA (2020b) Inventory of U.S. greenhouse gas emissions and sinks: 1990–2018. https://www.epa.gov/ghgemissions/inventory-us-greenhouse-gas-emissions-and-sinks-1990-2018. Accessed 30 June 2020

US EPA (2020c) Fast facts. https://www.epa.gov/sites/production/files/2020-04/documents/fastfacts-1990-2018.pdf. Accessed 30 June 2020

UNFCCC (2021a) Reporting requirements. https://unfccc.int/process-and-meetings/transparency-and-reporting/reporting-and-review-under-the-convention/greenhouse-gas-inventories-annex-i-parties/reporting-requirements. Accessed 29 Dec 2020

UNFCCC (2021b) National communication submissions from non-annex I parties. https://unfccc.int/non-annex-I-NCs. Accessed 29 Dec 2020

Chapter 6
Financing the Transition

6.1 Means for Financing the Transition

At the beginning of the decade of the 2020s, there is growing recognition of the need to boost investment in greenhouse gas mitigation. The United Nations Framework Convention on Climate Change (UNFCCC) is sponsoring a "race to zero" which aims to "rally leadership and support from businesses, cities, regions, investors for a healthy, resilient, zero carbon recovery that prevents future threats, creates decent jobs, and unlocks inclusive, sustainable growth" (UNFCCC 2021). Meanwhile, a task force established in 2020 called for scaling up voluntary carbon markets by a factor of 15 to meet corporate pledges to become carbon neutral or achieve "net-zero" objectives by mid-century (IIF 2021). In this chapter, we focus on the role that finance plays in advancing climate actions.

According to the Ricardian theory of international trade (Anonymous 2016), trade in comparable goods reflects the value of environmental damage which is assessed in three parts: direct damage, long-term damage (e.g., exhaustion of a resource), and intangible damage (loss of amenity) (Hetzel 1990). Raw materials are akin to natural capital. The role of financiers is to anticipate with sufficient certainty the flows of revenue (or loss) that this asset will generate in the future (Giraud 2021).

Monetary value is key to systems of finance. Monetary value can be represented by banknotes, bank deposits (checking or savings accounts managed within the banking system), and by computer networks for virtual money such as bitcoins. Currencies are instruments of exchange that represent an agreed-upon price for goods or services. Currencies only have meaning in the context of economic exchange. Currencies exist physically as banknotes and coins. "The first known coins date from 650 and 600 BCE in Asia Minor, where Lydia and Ionia's elites used silver and gold coins stamped to pay for the service of armies" (Kusimba 2017). For centuries if not millennia, the right to issue money and the duty to maintain its constant value has been a sovereign monopoly (Kusimba 2017).

Historically, banking activities were established in Europe to limit the risk of theft of physical currencies. A bank's "letter of change" replaced the need to travel with

© The Author(s), under exclusive license to Springer Nature Switzerland AG 2021
J. C. Shideler and J. Hetzel, *Introduction to Climate Change Management*,
Springer Climate, https://doi.org/10.1007/978-3-030-87918-1_6

pouches filled with gold and silver coins. Instead, a banker in one city asked his banking colleague in another city to pay an amount based on the letter. In general, in each place of commerce (medieval fairs in such cities as Troyes, in France,[1] for example), banks maintained accounts or a relationship with a locally established banking company. Banking activity in the Middle Ages was often reserved for non-Christians such as Jews or Arabs.

International trade has been key to the development of the financial sector. Indeed, trade was the traditional underpinning of finance. In recent years, however, the value of trade flows—imports and exports—has paled in comparison to international flows of capital. According to the most credible studies, financial flows (domestic and international) in 2007 were 73.5 times greater than nominal world gross domestic product (GDP). In 1990, this ratio was "only" 15.3 times (Gadrey 2014).

During the Middle Ages, trading in goods was controlled and taxes were paid at the entrance to each city ("dock dues" at island ports are a remnant of this tradition). With the passage of time, trade frontiers gradually assumed a national character, and under the authority of the World Trade Organization (WTO) and multilateral or bilateral treaties, customs taxes were reduced to facilitate international trade.

There are two types of currencies used in international trade: convertible currencies and non-convertible currencies. The convertibility of a currency depends on many factors, most prominently the economic strength of an economy or its political and economic acceptability in international trade. Thus, the Russian ruble only became fully convertible in 2006 (Millot 2006).

China's currency, the renminbi (RMB) or yuan (CNY)—literally "people's money"—was accepted from 1987 when China created the dim sum bond market, which trades bonds issued in RMB (Chiu 2020). The RMB is one of the top five currencies traded by SWIFT. SWIFT is the Society for Worldwide Interbank Financial Telecommunication, legally "S.W.I.F.T. SCRL," which provides a network that enables financial institutions around the world to send and receive financial transaction information in a safe, standardized, reliable, and secure format. SWIFT also sells software and services to financial institutions. Corporate identification codes (BICs, formerly "bank identification codes") are popularly known as "SWIFT codes." If a currency is not accepted internationally, due to the risk of default or other reasons such as international sanctions or the absence of relationships, international transactions can be settled in goods (oil, metals, or agricultural products), gold, or currencies such as the US dollar. On the foreign exchange market, the dollar reigns. About 90% of foreign exchange trading is transacted in US dollars.

---

[1] The weight of gold coins is still to this day measured in "troy" ounces, reflecting the system of weights and measures adopted by this French trading center in the Middle Ages. A troy ounce is 31.1 g compared to an avoirdupois ounce of 28.35 g.

6.1.1 Cryptocurrencies or Cashless System

For many years, some consumer advocates promoted decentralized money to strengthen the local exchange of goods and services. Some communities have issued coupons for this purpose (the French cities of Nantes and Grenoble, for example). Such local "currencies" are limited by the number of users willing to exchange of goods and services (Build Green 2019). These local mediums of exchange are different from the decentralized cryptocurrencies that are supposed to substitute for the established banking system. The latter rely instead on networks of computers that ensure the uniqueness of each transaction in a given cryptocurrency.

A cryptocurrency is "an unregulated type of digital money, which is issued and generally controlled by its developers, and used and accepted among members of a specific virtual community" (European Central Bank 2012). It relies on cryptography to secure and verify transactions as well as to control the creation of new cryptocurrency units (Kelas 2021). Figure 6.1 explains the "blockchain" process that makes this possible. An entry, once confirmed and modified by the issuers' ledger, is irreversible (Narayanan 2015) (European Parliament 2019).

To buy a cryptocurrency, a purchaser must be accepted as a member of the community and find a new cryptocurrency created as shown in Fig. 6.1.

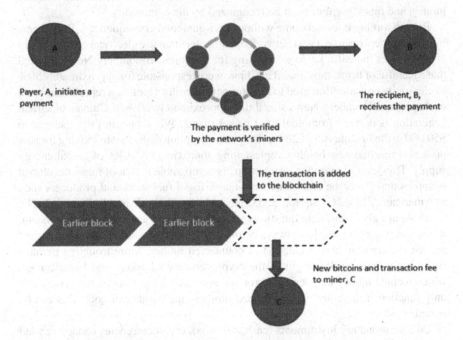

Payer, A, initiates a payment

The recipient, B, receives the payment

The payment is verified by the network's miners

The transaction is added to the blockchain

Earlier block   Earlier block

New bitcoins and transaction fee to miner, C

Fig. 6.1 Overview of the blockchain of a cryptocurrency (Söderberg 2018)

The price paid by A includes two components:

1. Payment to B
2. The remuneration of C (the "miner"), who found the cryptocurrency and certi-
    fied its unique number and guarantees its existence in the currency exchanged
    (bitcoin, Ethereum, Cardona, etc.).

The cryptocurrency listing is the result of the brokers' observation of the exchange
value, that is, the value that buyers are willing to pay for a good, here the cryptocurrency. Cryptocurrencies are created by individuals independent of a state who link
their computers to create a currency whose exchange value they recognize. The
"blockchain" corresponds to all computers that recognize that the algorithm that
is the medium of money is both unique and exchangeable. This is a value recognized only by the members of the network, who include the "miners" who create
the currency and who accept themselves as the creators of money. The purchaser of
a cryptocurrency, as described above, pays a price with two components, the sale
price accepted by the seller and the remuneration of the "miner." By contract, the
miners agree not to exceed the agreed limits on generation of the currency (21 million
units for bitcoin) (Crypto-France 2018). The bitcoin buyer becomes a member of the
network in that he can buy and sell his bitcoins, he does not by this status become
a "miner." For that he must have the ability to create the algorithm needed to create
money, and this algorithm must be recognized by the community.

Bitcoin mining does not come without environmental consequences. To find new
bitcoins, miners operate a vast network of computers that use electrical energy. When
bitcoin prices increase, so does searching for new ones. Bloomberg News reported
that a handful of firms, most based in China, were responsible for approximately 50%
of electricity consumption used to look for new bitcoins. Quoting a report from Bank
of America, Bloomberg News stated that approximately 60% of China's electrical
generation is derived from coal-fired power plants. When bitcoin prices surged to
$50,000 in the first quarter of 2021, energy consumption dedicated to looking for new
ones also reached new heights, representing approximately 0.4% of global energy
supply. This level of energy consumption "is comparable to that of many developed
countries and rivals the emissions from major fossil fuel users and producers such
as American Airlines Group Inc. and ConocoPhillips" (Hajric and Ballentine 2021).

There are also legitimate questions about the security and ethics of the system,
as the creation of money has always belonged to sovereign authorities, who protect
against the creation and circulation of counterfeit money. More troubling perhaps
is the sometimes-murky origin of the cryptocurrency's backers and beneficiaries,
which include money laundering, terrorism, and trafficking. Finally, the system can
only function if there are clearly defined limits to the number of coins that can be
issued.

Like all monetary instruments (cash, deposits), cryptocurrencies cannot be used
for green finance, now focused on initial investments for projects, assets, and
supporting expenditures as a green debt instrument. Money or deposits are used
in the short term; debts and investments have medium- and long-term uses. Today,

cryptocurrency is used notably for the laundering of money recycled from criminal activities or as a speculative tool such as gold or any other nonmonetary asset.

The distributed ledger technology of crypto money (the "blockchain"), studied today by the Central Bank of China, could in the future replace cash. Each individual in such a system would be assigned an account at the central bank. Enrolled persons could pay for goods and services using their mobile phones to access the blockchain without a bank transaction. This could be a good solution in developing countries to alleviate the problem of "unbanked" individuals (Piketty 2021).

6.2   Role of Finance in Transitioning to a Low-Carbon Economy

We have seen the emergence of many new financial products such as green bonds, climate bonds, and social bonds, which we will discuss in this chapter. A bond is a financial instrument, called an obligation or asset by financiers. It is a debt with a unit of account often valued at hundreds of million dollars, or even $1 billion, issued with a predetermined rate for a defined term by a company or a state and sold by banks to investors. The issuer agrees to pay a fixed amount of interest and reimburse the capital at maturity. A bond is generally part of a portfolio with other investment instruments.

The heart of the debate around climate change at the beginning of the 2020s relates to the allocation of capital. Capital investment is needed to mitigate climate change, to transition to a low-carbon economy, and to adapt societies to climate-related changes that cannot be abated or reversed. An alternate phrasing of this thesis is to define the conditions for sustainable finance, that is, financing that contributes to sustainable outcomes. Observers have not failed to note that there is a definite gap between markets and their exponential valuations and daily economic realities. We will talk in turn about the concepts of accounting and asset valuation according to the principles of corporate accounting, the concepts of returns on invested capital, the impact of capital on employment and its sustainability.

Key Definitions

Finance is a global term used to cover the activities of investment banks, large companies, and others to manage trade and transfers in equities, debts, deposits, and foreign currencies to support commercial and investment activities.

Direct access to financial markets is limited to large companies, wealthy people, pension funds, and insurance companies. Individuals can buy shares in the stock market or use banks as a proxy. Some financial activities are completed in nanoseconds between two computers, but other transactions by governments and companies involve much longer time frames such as very long-term loans to finance infrastructure development.

Capital is categorized by type. Seed capital is provided by founding shareholders to start a company or activity. This capital can be provided in exchange for newly subscribed shares through the transfer of financial reserves or contributions of tangible or intangible property. Many companies developing new concepts or processes acquire additional capital in stages. Venture capitalists may provide "early stage" series A and series B investments as a new company achieves milestones in demonstrating the value of its concepts. This is followed by a "syndication" stage when the company issues additional shares to public investors and others. A company's total capitalization is its value as determined by the stock market, which can rise or fall with each trade. In theory, stock market capitalization represents the total value of tangible and intangible assets less the debts of the company.

Investments. To develop an activity, a company needs buildings, machines, and utilities. Capital expenditures—CAPEX—are financed from capital or debt to pay for the property, plant, and equipment (PP&E) needed to conduct business activities. Equipment purchases are amortized over a period of a specified number of years (following accounting rules) until their cost is fully recognized on the company's balance sheet. Equipment typically is amortized over a period of five to ten years; buildings are usually considered to have a useful life of 40 years. Salaries, rent, and utilities are paid for through operational expenditures—OPEX. This means the cost is covered by cash flow in the ordinary course of business over a short term (usually 30 to 90 days in accordance with contractual terms agreed with suppliers).

6.2.1 What is the Value of a Traded Asset?

The question leads to the financial valuation of assets. This chapter is geared toward financing that has positive environmental benefits (the shorthand descriptor is "green"). It is a matter of supplying financial products, described as green, to meet the demands of investors, who wish to invest capital, for their own account or that of their clients, in instruments that will provide positive financial returns net of inflation in "green" projects, assets, or supporting expenditures. Financial performance can be an unbelievably bad indicator, but it is a necessary threshold for private investment (Ekeland 2021). As G. Giraud writes luminously: "For any resource to be matched with capital, it is necessary to be able to anticipate with sufficient certainty the future flow of income (or losses) it will generate" (Giraud 2021).

Accounting data is an essential input to identify the value of a business, and it is the essential work of analysts to dig into published accounting documents and supporting information when it exists. In many groups, especially family groups, transparent information is not publicly accessible. We agree with the definition of accounting by F. Dejean, Professor at Paris Dauphine University: "Frequently perceived as a neutral technique, accounting is, on the contrary, a subjective discipline that works silently…, which shapes the conceptions of the staff of companies…. It is not a photograph, but an image developed according to certain assumptions" (J. Richard quoted in Dejean 2021, p 71). Accounting is used to measure profit and determine the amount of dividends distributed to shareholders (Dejean 2021). Accounting is based on the precautionary principle, hence the use of the historical cost basis to assess the value of assets and limitations on the ability of managers to distribute fictitious profits.

This principle was abandoned in 1973 when an international standard-setting body, the IASC, now IASB, created an accounting rule that with the stroke of a pen called for physical assets to be valued at their market value. This had the effect in most cases of adding value to the balance sheets of companies and effectively stimulating the distribution of fictitious gains (Dejean 2021). These practices led to Enron's designation as "the Best of the Best" by Goldman Sachs Bank and the disappearance of the "big five" accounting firm Arthur Andersen (Snégaroff 2015).

Prudential accounting (practiced by many unlisted companies) assesses an asset at its acquisition price, less its amortization which allows it to be replaced at the end of its useful life. Such assets must produce goods and services to pay salaries, rents, and all related expenses including taxes and produce a return on capital for shareholders. The valuation of a business results from the difference between assets after depreciation, less debts, on resale. It is calculated based on the expected profit for the capital invested by the buyer. In this accounting, it is a question of evaluating the company in years of future profits to repay the borrowed capital and pay a return on invested capital.

In financial accounting, the purpose is to assess the asset as a representative share of the capital, based on the company's stock price, i.e., the amount of capital multiplied by a factor that represents the market's estimated value on the closing date. For example, according to the Washington Center for Equitable Growth, Facebook's intangible value without market values would be equal to one-sixth its current capitalization (Leiserson 2017). The remaining five-sixths represents intangible value called "goodwill"—an accounting term that represents the difference between the value of a company's assets and its stock market valuation. As a result, at the initial public offering (IPO) it is not necessary to make a profit to have a very high value based on goodwill. The pharmaceutical company Moderna and maker of a covid-19 vaccine is a case in point.

In 2020, the company's stock price rose from $43 to $189 even before turning its first quarterly profit in 2021. It was Moderna's goodwill value and expectations of future profits that raised the price. If this situation seems perfectly legitimate for a research and development company like Moderna, it is more problematic for a company whose market is mature and subject to disruption, as is the case for "bricks and mortar" retail stores confronting the rise of digital competitors.

A. Berger writes that listed banks must meet the financial markets' demand for profit margins: around 11–12% currently, even though the so-called reasonable reference in financial circles is 15%. Unlisted banks such as local or cooperative banks, which still exist in the real economy, can be satisfied with profit margins of 5–6%. This difference in profitability completely changes the face of the green economy resulting from the energy transition (Berger 2021).

Maintaining the stock market price of listed assets is a key criterion in valuing executive careers and income. When the only measured performance is that of financial valuation, the result is what economists call the "financialization of the economy." Investment is replaced by a narrow focus on the price of shares on the stock market which must be maintained to satisfy shareholders and provide them with capital gains. The real economic value of the enterprise is left out of the equation.

Companies often use profits to buy back shares to maintain the shares' stock market price. Before the onset of the pandemic-induced health crisis, share buybacks increased in Europe, where they accounted for 32% of the amount distributed to shareholders (by mid-2019 according to Morgan Stanley Bank). This amount has room to increase further as it is far behind that of the US markets (68% of the amounts distributed) (Boursorama 2020).

A. Berger sheds interesting light on the rate of return on invested capital. He gives the example of French highways, which were sold to financial groups on the basis of an assessment by the French Transport Regulatory Authority that the average return on invested capital should be between 4 and 5.6% after allowing for sustainable highway management. In reality, the financial groups that privatized French highways generated returns between 9 and 11%, reaching 24% in 2014, enabling them to distribute dividends three times higher than those distributed by motorway companies before privatization. The average profit was €880 million before privatization (2003–2005) which increased to €1.6 billion on average for the period 2006–2013. The dividend distributed for the first period was €500 million (56%) to €2 billion or 125% (Berger 2021, pp 27–31). The private operators starved the motorways of investment in order to distribute dividends. If we are committed to ensuring that investments in green projects, assets and supporting expenditures are a powerful mechanism for financing environmentally beneficial assets, finance of this type cannot be the model.

To be productive, green investments should not generate high financial returns in the short term, because:

– It is a brake on environmentally friendly investment.
– Profitability is long-term.

Otherwise, the financial market creates outsized economic rents on new, less polluting, and low-GHG products. We have the case of hydrogen, whose development will be carried out using policy incentives, as current techniques are up to 80% more expensive than other renewable energies. In fact, investors demand the same 15% returns as for mature investments, even if they result in prohibitively high selling prices in the market. In most mixed financial systems, long-term financing is covered by state-guaranteed savings. In some respects, green finance is competing with hedge funds for investors' available capital, leading to a call for regulations to limit these flows (Grandjean 2021).

NGOs use the term "name and shame" to refer to calling out false statements such as "compliant with less than 2 °C" or "in line with the Paris Agreement" (Clerc 2021) when the plans set by individual companies or investment portfolio managers fall short of the actions needed to achieve compliance with the Paris Agreement's less than 2 °C goals (Parance 2021).

At a minimum, corporations have a duty of care relating to the non-harmful nature of the financed project which should be reflected in their ESG procedures and on-the-ground practices (Parance 2021). A groundbreaking ruling by a district court in The Hague (Netherlands) in May 2021 cited Royal Dutch Shell's duty of care obligation regarding the human rights of those affected by climate change (Raval 2021). If the ruling withstands appeal, it could mark the beginning of a process where major contributors to the emissions that cause climate change are held to account.

6.3 How Does Green Finance Work?

Figure 6.2 from Dutch researchers depicts roles and responsibilities in a program to

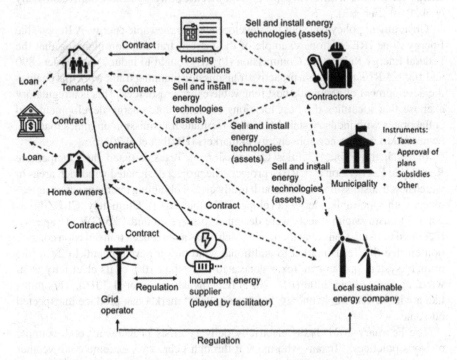

Fig. 6.2 How to deliver renewable energy in the housing sector (Bekebrede et al 2018)

deliver low-carbon electricity to residential homeowners

The example is based on a European model but is applicable anywhere. Regulation provides a framework for deploying renewable energy to the housing sector. Energy is a strategic utility in all countries, and regulation plays a key role. The US National Renewable Energy Laboratory (NREL) concluded that "Rapidly declining costs of wind and solar energy technologies, increasing concerns about the environmental and climate change impacts of fossil fuels, and sustained investment in renewable energy projects all point to a not-so-distant future in which renewable energy plays a pivotal role in the electric power system of the 21st century" (Chernyakhovskiy et al 2016).

In the USA, as in other major developed countries, laws and regulations governing the electric power sector and renewable energy interconnections have evolved significantly in recent decades through a process known as restructuring. The term "restructuring" can refer to two different parts of the utility market: wholesale and retail electricity markets. Restructured wholesale markets comprise a range of different market-based approaches to balancing bulk power supply and demand while creating an institutional separation between transmission operations and generation. Retail restructuring refers to the introduction of consumer choice among competing suppliers of end-use electricity services. Interconnection policies have gradually evolved to expand participation opportunities for independent generators, including renewable energy generators, in different aspects of the electric power system (Chernyakhovskiy et al 2016) (Fig. 6.3).

Government policy can spur the development of renewable energy. A Renewable Energy Zone (REZ) is one example of the kind of transmission planning that the Federal Energy Regulatory Commission (FERC) sought to induce with Orders 890 and 1000. A REZ process is a proactive transmission planning framework that enables the development of a region's best renewable energy resources. This is a regulatory exercise that identifies the best locations for renewable energy development and collaborates with industry stakeholders to facilitate transmission upgrades that may be needed to bring renewable energy to markets (Hurlbut et al 2016).

In 2005, the Public Utilities Commission of Texas initiated the Competitive Renewable Energy Zones (CREZ) project. The project designated wind-rich areas in which new transmission would be built in advance of obtaining interconnection agreements with renewable energy developers (Hurlbut 2010). Ultimately, CREZ high-voltage transmission projects were designed to serve around 18.5 GW of capacity. Increased transmission capacity in wind-rich areas and reduced transmission congestion opened the Texas market to additional wind power development. In 2014, the primary system operator in Texas generated more than 10% of its electricity from wind, up from 6% in 2006 (EIA 2021) (Chernyakhovskiy et al 2016). This looks like a win for renewable energy, but a deregulated market can produce unexpected outcomes.

The February 2021 Texas electricity delivery crisis provides a good example of worst practices. "Texans who made it through February's extreme cold weather without losing power or natural gas must have felt lucky. But for some, keeping their electricity through the blackout may turn out to be more traumatic than losing it. An

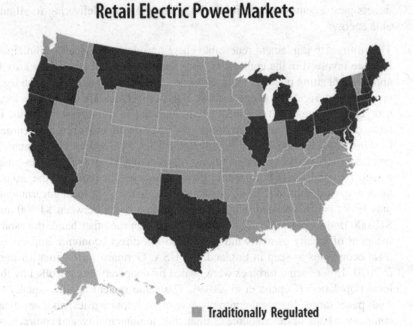

Retail Electric Power Markets

Traditionally Regulated
Competitive

Fig. 6.3 Regulated and deregulated US energy market (US EPA 2021)

undetermined number of homeowners have been shocked to receive bills running into the thousands of dollars—in some cases, over $15,000 for a month's worth of power. As someone who has spent the past two decades studying electricity deregulation, I know that extreme power bills in Texas result partly from the state's market-driven approach to running the power grid. But decisions by state regulators also had a hand. Measures that were originally intended to give logical signals to the electricity market and encourage conservation during hot spells were not up to the task of managing this cold-weather crisis" (Blumsack 2021).

Along with 16 other states, Texas has deregulated its power generation market. The Texas market has a wholesale and a retail component, like the markets for many other goods. In the wholesale power market, companies that generate electricity compete with one another to provide power on a market operated by the Electric Reliability Council of Texas, or ERCOT. In the retail market, electricity retailers buy power wholesale from ERCOT, add transmission and distribution charges to the wholesale generation cost, and resell that electricity to households and businesses. These resellers include Texas' five electric utilities, which offer fixed and regulated prices in the areas of the state that they serve. Hundreds of others, known as retail providers in the Texas system, are unregulated and can offer electricity to consumers at any terms and at any price.

Contracts play a central role in producing, managing, and delivering fossil and renewable energy:

(1)   The contract to implement renewable energy production. Usually, municipalities are involved in the contract as taxing authorities, but potentially also as approvers of siting plans. In many countries, community boards have the legal or customary rights to approve siting decisions. Taxes are one of the local sources of revenue that are most often analyzed in this literature. Thus, in some countries, local administrations levy taxes on the gross revenues earned by wind farms located in their jurisdictions. This is the case of Greek municipalities which receive annual payments representing 3% of the gross revenues earned by wind farms built within their boundaries; in Portugal, the annual amount corresponds to 2.5% of gross revenues. In the USA local governments may levy taxes associated with wind power in a range between $4,000 and $12,000 dollars per MW (Slattery et al 2012). On the other hand, the establishment of locally owned wind farms also has direct economic impacts on local economies as seen in England, the USA, Germany, or Denmark where in 2010, 15% of wind turbines were owned by cooperatives controlled by the local population (Copena et al. 2019). The same system applies equally to solar panel farms. Standard contracts may involve four parties: an investment company, a bank or other source of financing, a municipality, and contractors. Production is measured by the quantity of electricity delivered for consumption or storage.

(2)   The contract to deliver electricity by incumbent energy suppliers, who play a role in managing the grid network. Electricity cannot be easily stored, so grid operators balance loads according to the availability of renewable and nonrenewable generation sources (taking into account baseload sources such as low-carbon nuclear production and gas turbine plants for peak hours of demand). Renewable electricity generation is supported by a mix of conventional energy sources. Grid balancing decisions generally are made on the basis of offered price, but carbon emissions should be considered as well. Decentralized production, such as from rooftop solar panels, can satisfy some domestic demand, with high-performance batteries supplying electricity for hot water and heating.

(3)   Contracts for renewable energy are supposed to support fulfillment of requirements and the measurement of environmental performance. In the solar example, the primary indicator of output is the kWh, meaning the quantity of electricity produced per hour. Another indicator is the amount of surface area, in square meters, covered by solar panels. For solar generation, an efficiency metric in kWh/m^2 expresses the quantity of delivered renewable energy from a specific installation. The power of solar photovoltaic panels is expressed in Watt peak, abbreviated as Wp. The number of cells in the panel and their quality defines the power of a given panel. The current power standard for photovoltaic solar panels is around 300 Wp. Optimal peak power is achieved under the following conditions (Dualsun 2021):

- A temperature of 25 °C which is neither too cold nor too hot
- Incident sunshine of 1000 watts per m²
- An inclined surface of 30 °C and directed toward the south (in the northern hemisphere)
- Clear sky.

As Fig. 6.2 indicates, finance also plays a central role:

(1) Contractors use finance (loans or bonds) to develop housing, directly or indirectly for individuals and housing authorities, as well as commercial centers which sell products and services to consumers.
(2) Investors, or banks as proxies for investors, finance energy projects which pay contractors to design, implement, and manage energy systems.
(3) Finance makes possible the purchase of electricity-consuming appliances in homes.
(4) Multiple regional networks in the US transmit electricity and balance generation and consumption across their grids; US electricity final demand is 4 TWh and represents 17% of total world demand (EIA 2020).

If we wish to summarize finance, we have two types of actors:

(1) investors with available funds such as cash or bank deposits
(2) debtors who need funds for projects, assets, or supporting expenditures.

6.4 Who are the Finance Sector Actors?

The concept of a financial system remains highly abstract for nonspecialists. The reality is shifting, as the financial system covers widely different situations:

- Countries, regions, and multilateral institutions that all operate within specific financial systems
- International mechanisms for international trade
- Nanosecond transaction speeds from one market to another using fiber-optic cables and secure connections; in the 1970s, one bank check in the amount of $1 billion was sent by courier on an Air France Concord flight across the Atlantic to be presented for payment.

6.4.1 International Financial Institutions

The International Monetary Fund (IMF) was created in 1944 as part of the Bretton Woods agreement. Bretton Woods helped reestablish international economic exchanges and limit economic disruption after World War II.

With the support of the World Bank's financial resources, the IMF advises countries in case of a monetary crisis and international payment defaults. It is well known to push countries to implement debt restructuring plans. The IMF and World Bank, more formally the International Bank for Reconstruction and Development, played a role to limit the effects of the monetary crisis of 2008 by providing development assistance.

The World Bank represents the central banks of each country. It delivers funds to countries through loans or grants and may apply nonfinancial criteria to finance the countries based on sustainability scenarios. Since 2015, the United Nations has promoted an international environmental and sustainability consensus through its Sustainable Development Goals (SDGs).

NGOs have frequently criticized the International Finance Corporation (IFC), a member of the World Bank Group, for not being able to track its money due to its use of financial intermediaries. For example, a report by Oxfam International and other NGOs in 2015, "The Suffering of Others," argued that the IFC was not performing sufficient due diligence or managing risk in many of its investments in third-party lenders. Other criticism focused on the IFC's excessive lending to large companies or wealthy individuals who could already finance their investments without help from public institutions such as the IFC, and the inadequate positive development impact of such investments. An example often cited by NGOs and critical journalists was the IFC granting financing to a Saudi prince for development of a five-star hotel in Ghana (Geary 2015).

6.4.2 National Banking System

Each country is organized on the same principle: its central bank can create monetary instruments (cash, loans) and regulate banking activities. Within a country, "retail" commercial banks are variously organized as private companies or cooperative associations. Some are based on sectarian principles (Christian and Islamic are the most prominent). They receive local, national, or international funds from depositors which they use to make loans (short-, medium- or long-term).

Originally, the banking system was based on the confidence of two bankers to complete transactions in different locations and to limit risks associated with the transport of money. Now international networks move money via the Internet from one country to another instantly. This system supports international trade.

Lending is regulated to limit the risk of inflation, which can be caused by an increase of the amount of money in circulation. Banking deposits are generally invested in short-term financial instruments, and banks create money by using leverage to offer medium- or long-term loans. Banks receive deposits which they in turn put to work as loans and investments, maintaining a percentage as reserve capital to cover any demand for deposit reimbursement.

In a financial crisis, central banks can lend money to national or local banks to limit the risk of insolvency. The US Glass–Steagall Act in 1933 created a system of deposit insurance to guarantee deposits. In 1933, the insurance covered deposits up to $2500, an amount that has now grown to $250,000. In parallel, commercial banks and investment banks were separated. After the 2008 crisis, the same system for deposit insurance was created in Europe. Since then, commercial bank accounts have enjoyed a state guarantee of €100,000 per account and household, and bank capital reserve requirements were raised under the regulatory framework called Basel III.

The systematic risk of financial institution collapse was not fully contained. In 2019, the largest European bank, Deutsche Bank in Germany, was obliged to reduce its personnel by 20% and transfer nonperforming loans to a new "bad bank" to avoid collapse and limit the risk of a systemic banking crisis in Europe. The bad bank received €288 billion of assets that Deutsche Bank management had written down to €74 billion (a loss of 75%) (Bloomberg 2019).

In Europe, some banks are global (investment and commercial) such as BNP Paribas. Banks of this type enable clients to access the financial markets and sophisticated market products used for hedging or speculation. In the past, the two banking activities could not be combined in the same banking company. Global banks undertake varied commercial operations—banking, insurance, real estate—as intermediaries. They also develop derivative products based on indexes and options. Derivatives are an outgrowth of securitization, a concept that we will return to later. Some economists think European states are not able to cover the risk of non-payment by global banks (Giraud 2012), the deposit guarantee amounts being larger than the states' annual real gross domestic product (GDP).

In Europe, commercial banks finance 70% of economic activity. Financial markets meet the remaining demand. Investment banks provide access to funds in financial centers around the world, typically for large companies operating in international markets or for large private fortunes. Investment banks and private banks provide wealth management services, generally serving accounts containing a minimum of $1 million in liquidities. Climate change activities are considered in the banking market as a niche market to finance specific products. This niche attracted greater attention in the early 2020s as evidenced by reporting in the Financial Times' headline "Wall Street's new mantra: green is good" or in the Wall Street Journal's headline "Climate change emerges as a compliance focus for SEC" (Michaels 2021).

6.4.3 Companies' Financial Services

Large companies manage treasury flows and maintain substantial funds to cover investments in industrial or other assets. They provide short-term cash management and cover currency and purchasing risks directly in the respective markets: raw materials markets, oil markets, carbon trading markets, etc. In a period of massive disinvestment (2018–2019), financial flows can increase profitability and encourage speculation. Some of the most important decisions that company managers make are about the allocation of resources. Do they use them to finance external growth or purchase assets? Or do they buy back shares to bolster the company's stock price? Do they borrow money by issuing corporate bonds or finance activities through bank loans? "Wise" investment decisions can lead to medium- to long-term growth and profitability, while "ill-considered" decisions can overburden a company with debt and lead to financial underperformance or even bankruptcy.

6.4.4   Sovereign Wealth Funds (SWF)

Sovereign wealth funds were developed under the influence of petrodollars to manage the influx of oil revenues (Qatar, Saudi Arabia) and to permit the Union of Soviet Socialist Republics (USSR) to access international payment systems during the Cold War. Contemporary SWFs manage investments made by different states, such as the French Public Investment Bank (BPI), the Norwegian oil fund which manages North Sea oil and gas revenues. The Kuwait Investment Bank created the first SWF in 1953. They aim to invest over the long term to build reserves that can be used in the future once revenues diminish from the exploitation of oil and other raw materials. Large funds are able to apply state policies such as investment bans on coal energy producers (Holter 2016).

6.4.5   Pension Funds

Contractual (private) or governmental pension funds manage investments and distribute pensions to eligible persons. Preliminary data for 2015 show that aggregated pension fund investments in Organization of Economic Cooperation and Development (OECD) member countries amounted to $24.8 trillion (OECD 2016). However, when pension fund investments are expressed in national currency, they grew in all countries in 2015, except for Denmark, Luxembourg, Poland, and the USA. The overall decline in total OECD pension fund investments is largely driven by the decline in the USA, which represented 58% of all pension fund investments in the OECD area. In addition, the growth rate of pension fund investments was below 2% in Japan, the Netherlands, and the UK, representing altogether a further 22% of investments within the OECD area (OECD 2016).

The US federal Old-Age and Survivors Insurance Trust Fund is the world's largest public pension fund which oversees $2.72 trillion in assets. The assets held by the Trust Fund are US government securities. No problem existed as long as the US Social Security Administration could cover its monthly pension check obligations from current payroll tax collections. Starting in the 2020s, however, payments from the Trust Fund were projected to exceed income, meaning that the Treasury needs to borrow money to compensate for the reduced government revenue available for other government programs (Pattison 2015). The situation is typical of all developed countries that have a pay-as-you-go system that uses current revenues to pay retiree pensions that are based on prior contributions and promised benefits.

6.5 A Quick Survey of the Finance Sector

When people speak of "finance," they generally think in terms of billions of US dollars as a minimum. In 2020, cumulative green bond issuances amounted to around $1 trillion (Jones 2020). Green bonds amounted to about 5% of total bond issuances, a substantial amount but still a fraction of the total amount financed ($74 trillion in 2020).

Following the Paris Agreement in 2015, the UNFCCC and the World Bank commissioned a study to document climate-related investment flows by countries (Climate Policy Initiative 2018). Figure 6.4 maps by source and intermediaries world-wide global commitments in 2017–2018 to climate change investments (Macquarie et al 2019).

Different points must be highlighted:

- The global amount of $574 billion as an average annual amount financed in 2017–2018 seems plausible if it is compared to other sources, such as the issuance of $389 billion worth of green bonds (Climate Bonds Initiative 2018)
- 52% of committed funds are public or multilateral; states play a major role in financing climate policies.
- 48% of reported finance came from the private sector, distributed among short-term issues (commercial paper, 17%) and balance sheet financing (56% with equity and 44% by debt).
- 66% was financed by debt instruments (public and private).
- The primary targets for investment were renewable energy and efficiency (65%) and low-carbon transport (24%).

Financial instruments employed included the following:

- $27 billion in grants delivered by multilateral institutions or bilateral agreements
- $65 billion in low-cost projects to help developing countries finance low interest rate loans (some developing countries may have refinanced earlier revolving credit facilities)
- $219 billion project-level market-rate debt (green bonds and green loans)
- $24 billion project-level equity investments in the capital of a specific company or a subsidiary
- $217 billion to finance housing and commercial buildings with equity or debts (56% with equity and 44% by debt).

The contribution of equity to the formation of investment needs to be discussed. When someone buys shares on the market, the buyer's transaction has zero effect on investments (green or not). The buyer is simply replacing an existing stockholder. New shares created during a capital increase are rarely used for material or intangible investment, but they increase equity, which can be used for investment. In addition, the new shares are created for the entire life of the company, which typically does not correspond to the life of any particular investment. In a comprehensive analysis

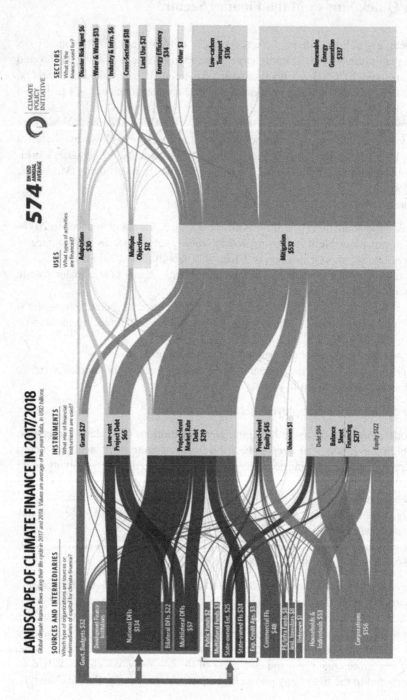

Fig. 6.4 Climate change investments in 2017–2018 in billion USD. *Source* Climate policy initiative © 2020 (Macquarie et al 2019).

of a portfolio, only capital increases should be considered relevant to investments, on the condition that the capital increases:

– Are designated for a specific use and
– Represent undiluted capital.

6.6 Multilateral Banking and Finance

No discussion of green finance would be complete without reference to the multi-national development banks. The multilateral development banks were among the earliest developers of taxonomies. Following the issuance of the first named green bond by the World Bank in 2009, other multilateral development banks helped transform the precedent into a movement. Important early work was done by the European Investment Bank. Subsequently, the Inter-American Development Bank, the Asian Development Bank and others agreed in 2011 to a joint framework for tracking and reporting climate change mitigation and adaptation finance.

Multilateral banks and other development finance institutions have contributed financing actions to combat the climate crisis, even if their financial firepower alone is insufficient to meet the "trillions of dollars" in financing needs that the world requires (The World Bank 2021).

The initial funding of the Green Climate Fund, a facility established by the UNFCCC, highlights the fact that public resources are too limited to fully finance decarbonization and adaptation (Green Climate Fund 2019). In parallel, private capital sits on a large quantity of cash in short-term accounts that could be put to work financing climate change mitigation and adaptation. "Green finance" has the potential to become as significant for the twenty-first century as was the development of industry in the nineteenth and the first exploration of space in the twentieth.

A role for financial instruments was designed into the Kyoto Protocol with its flexible mechanisms. CDM credits were key to financing early mitigation investments and regulatory cap-and-trade markets featured trading of both emissions allowances and offset credits. We described the mixed results of these market mechanisms—primarily due to the overallocation by regulators of emission allowances and the low price of offset credits in the voluntary markets—in Chap. 2.

6.7 Development Aid

Many countries offer assistance to developing countries to help them meet their economic, social, and environmental needs. Distribution of aid of this kind may be through specialized development agencies such as the US Agency for International Development or the Agence Française de Développement. In addition to development agencies, some countries have established specialized development banks to provide development assistance. Development banks may be national, bilateral, or multilateral. The capital of these banks usually is underwritten by contributions from national governments. As shown in Fig. 6.3, development finance institutions are major sources of both low-cost and market-rate project debt for climate change mitigation and adaptation initiatives. From 2020, their contributions are expected to form a significant part of the $100 billion per year in funding that developed countries pledged in 2009 at COP 15 in Copenhagen and reaffirmed in the Paris Agreement. These funds are intended for developing countries to help them cover their climate change mitigation and adaptation needs.

Bilateral development aid amounted in 2016 to $142 billion (OECD 2017) including debt relief. As the UN considers that it is appropriate for countries to spend 0.7% of gross domestic product (GDP) on development aid, international statistics show that apart from a few northern European countries (Denmark, Luxembourg, Norway, Sweden and the UK) that meet this objective, the current global average level is 0.29% (USA 0.19%, France 0.36%). More needs to be done on the part of developed countries to meet existing commitments.

6.8   The Birth of Green Finance

The green bond market kicked off in 2007 with AAA-rated issuances from the World Bank and the European Investment Bank (EIB). The first green municipal bond was issued by Massachusetts in June 2013. Gothenburg (Sweden) issued the first Green City bond in October 2013. US states are major green bond issuers, but issuers also include the Province of Ontario, the City of Johannesburg, and the Province of la Rioja (Argentina) among others. Local government green bonds continue to grow.

The wider bond market started to react after the first $1 billion green bond (offered by the International Finance Corporation (IFC) was fully subscribed within an hour of its offering in March 2013. Previous green bond issuances were denominated in millions of US dollars rather than billions.

The green bond market has seen strong growth, with the market starting to take off in 2014 when $37 billion was issued. In 2018, issuance reached $167.3 billion, setting yet another record. In November 2015, there was a turning point in the market when Vasakronan, a Swedish property company, issued the first corporate green bond. Large corporate issuers have since included SNCF, Berlin Hyp, Apple, Engie, EDF, ICBC, Crédit Agricole and others. In 2021, France Tresor, a part of the French treasury department, emitted €300 million in green OAT (fungible treasury bonds), in this case with a maturity of 23 years.

The green bond market encompasses a variety of financial instruments. SolarCity (now Tesla Energy) issued the first solar asset-backed security (ABS)[2] in November 2013 (Climate Bonds Initiative 2021). The biggest ABS issuer was Fannie Mae, one of two quasi-public residential mortgage lenders in the United States. Other green

---

[2] An asset-backed security pools multiple loans in a single security, sometimes with different "tranches" representing higher and lower levels of credit worthiness.

asset-backed securities soon proliferated. In addition to solar ABS, the market has so far been presented with green mortgage-backed securities (MBS), green residential mortgage-backed securities (RMBS), green commercial mortgage-backed securities (CMBS), property-assessed clean energy asset backed securities (PACE ABS), automobile ABS, and receivables ABS. The green issuance momentum has continued, with more than $500 billion in green bonds currently outstanding.

External review of green bond issuances is common practice but only recommended by the Green Bond Principles. The Climate Bond Standard and ISO 14030-4 mandate third-party verification, and many investors expect green-labeled debt instruments to be either third-party verified or second-party reviewed.

Climate bonds, designed to mitigate climate change or aid in climate change adaptation, account for almost half of green bonds. Another variant on the theme is "Blue Bonds" whose financing is intended to protect oceans and their fauna. This is the route the Seychelles (an island nation in the Indian Ocean) chose to protect its marine world which is crucial to the survival of its population and economy. Early projects, carried out with the support of the World Bank, financed sustainable fishing activities and the protection of marine areas. Similarly, the Netherlands has also issued Water Bonds to finance their adaptation to sea level rise.

Since 2017, sustainability bonds have emerged that address social and environmental impacts. A variant on this theme is "SDG bonds" whose use of proceeds is linked to the achievement of UN Sustainable Development Goals (SDGs). Bonds of this kind may finance a "just transition," considering changes in employment and the economy related to the ecological transition. The methodologies for sustainability and SDG bonds are still in their infancy, but their development is very rapid.

6.9 International Standards for Green Finance

Following the publication of Green Bond Principles by the International Capital Market Association and the emergence of the Climate Bonds Initiative (CBI), the International Organization for Standardization ("ISO") launched a new project to develop standards addressing green debt instruments.[3] ISO 14030 "Green debt instruments" is a standard in four parts:

- *ISO 14030-1 Green debt instruments—Process for green bonds*
- *ISO 14030-2 Green debt instruments—Process for green loans*
- *ISO 14030-3 Green debt instruments—Taxonomy*
- *ISO 14030-4 Green debt instruments—Verification programme requirements.*

The following figure explains the relationship among the four parts of ISO 14030 (Fig. 6.5).

---

[3] The standardization work was carried out in ISO Technical Committee 207 Subcommittee 4, chaired by John Shideler, who also convened the Working Group responsible for the standards.

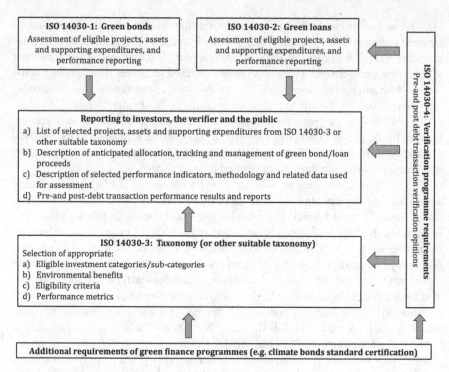

ISO 14030 aims to define how green debt instruments can help finance the $1 trillion of investments per year needed to achieve a low-carbon conversion of economies. Today, finance covers only about 60% of the identified demand (see discussion above).

Parts 1 and 2 of ISO 14030 define the process steps for green bonds and green loans, respectively. A third part, the taxonomy, categorizes economic activities by industrial sector. Users of parts 1 and 2 who find the activity they wish to finance in Part 3 can safely assume that the activity is green as long as it meets any defined thresholds and does not fall in the category of excluded activities. Part 4 defines verification program requirements for parties that wish to claim that their green debt instruments conform to the requirements of the ISO 14030 series of standards.

The taxonomy plays a key role in assessing whether the projects, assets, and supporting expenditures to be financed are environmentally beneficial ("green"), taking into account environmental impacts related not only to climate change but also other environmental aspects such as water, pollution, biodiversity and contribution to the circular economy. It also identifies adaptation approaches suitable for application in many economic sectors (ISO 2021c).

Users of parts 1 and 2 of ISO 14030 are provided the option to use Part 3 as their taxonomy or to select another taxonomy in its stead. This option was provided because regulatory bodies in many countries—such as the European Union, China, and India—have developed green or sustainable finance taxonomies for use in their jurisdictions, or had plans to do so (e.g., Canada). Moreover, reputable pioneers in green finance like the Climate Bonds Initiative have developed taxonomies through multistakeholder consultations that are widely accepted in the marketplace.

Taxonomies are key to the labeling of debt instruments as green. Investors have demonstrated a keen appetite to purchase green debt, with offerings of bonds often being oversubscribed by many multiples of their face value[4] (Climate Bonds Initiative 2017). On the other hand, investors sometimes shun bond offerings where the environmental benefits imputed to the issuance lack credibility. This occurred, for example, with a proposed green bond issuance by Repsol, the Spanish oil refiner. The bond was turned down for certification by the Climate Bonds Standard and excluded from a green index by Bloomberg. The criticism was that the proposed use of proceeds was "not green enough" as Repsol sought to finance incremental energy efficiency improvements at its refineries (Mullin 2017). The process for issuing a green bond or originating a green loan is intended to ensure the environmental integrity of the green debt instrument market.

6.9.1 The Process for Green Bonds

The objective of ISO 14030 Part 1 is to define a process for designating a bond as "green" that will assure investors that the financed projects and assets achieve positive environmental benefits that merit the label. The standard (ISO 14021a) achieves this by defining requirements for:

- The process of selecting and evaluating projects, assets, and supporting expenditures
- The management and use of funds
- Environmental impact assessment
- Reporting
- Validation and verification.

The first step in deciding whether to label a bond "green" is to consider the eligibility for this label of the proposed projects, assets, and supporting expenditures that will be financed. ISO 14030 Part 1 requires that all included projects and assets meet the eligibility criteria of either ISO 14030 Part 3 or another selected taxonomy. Moreover, the projects and assets shall "positively contribute to at least one environmental objective, which will be assessed and, where feasible, quantified by the issuer," and

---

[4] Investors are said to "oversubscribe" a bond issuance when there are more orders for the bond than bonds on offer for sale. In such cases, a bond issuer may choose to increase the financed amount to accommodate some of the demand or leave a portion of the purchase offers unfulfilled.

"manage environmental aspects to the extent practicable in order to avoid or mitigate associated negative environmental impacts." The environmental integrity of the debt issuance is demonstrated by evidence of conformity with these two requirements (ISO 2021a).

Many issuers of green debt define a "framework" for their green bonds that provides guidance to operational personnel in their organization as well as to investors. Framework documents can describe the organization's objectives for green finance and the procedures employed to ensure conformity with standards such as ISO 14030.

ISO 14030 Part 1 presents a list of seven environmental objectives that the issuer's financing may support[5]:

– Climate change mitigation
– Climate change adaptation
– Sustainable use and protection of water and marine resources
– Waste prevention and recycling
– Transition to a circular economy
– Control and prevention of pollution
– Protection and restoration of ecosystems.

This list of objectives is not exclusive, but indicative.

Part 1 foresees the possibility that an issuer will seek to fund projects and assets that are not mentioned in Part 3 or its selected taxonomy. In this case, Part 1 allows the issuer to include the projects and assets upon their its validation, which should be performed according to the principles and requirements of ISO 14064 Part 3:2019 Specification with guidance for the verification and validation of greenhouse gas statements.[6]

ISO 14030 Part 1 requires issuers to employ procedures for the management and use of funds raised through green bond issuances. These procedures ensure that the financing raised by the green bond is allocated exclusively to green projects, assets, and supporting expenditures and tracked. It also requires the issuer to be transparent about how any unallocated proceeds are placed and recommends that they be conservatively managed and not mingled with funds that are intended for uses inconsistent with those for a green bond.

Issuers of green bonds are required to evaluate and track material environmental aspects and risks associated with eligible projects and assets. They shall also define environmental performance indicators and track significant positive and negative environmental impacts. The required reporting highlights how investments direct or redirect human activities and provide quantitative or qualitative information

---

[5] Objectives ©ISO. This material is reproduced from ISO/FDIS 14030-1:2021, with permission of the American National Standards Institute (ANSI) on behalf of the International Organization for Standardization. All rights reserved.

[6] While this standard focuses on the verification and validation of greenhouses gas statements, it has been used more broadly as a framework for verifying and validating other subject matters, such as organization statements about water consumption and management, sustainability reports, and green debt instruments.

about the beneficial outcomes achieved with green investments. For climate change-related indicators, measured outcomes are primarily expressed in terms of the carbon footprint of projects and assets.

Part 1 defines requirement for reporting to investors both before the issuance of the bond and after its issuance. This includes reporting on the impacts achieved through the financing of green projects, assets, and supporting expenditures. The standard requires information reported to investors prior to bond issuance to be verified.

6.9.2 The Process for Green Loans

ISO 14030 Part 2 (ISO 2021b) defines requirements for green loans. Its requirements are similar to those in Part 1 with one major exception. The loan market supports both corporate borrowers and pools of borrowers who may seek financing for a specified purpose such as installation of roof-top solar panels on residential properties. Loans are contracts that involve advancing money to borrowers in exchange for an agreement to repay with interest on a periodic basis. Banks dominate the lending market but loans may also be made by other financial institutions as well as private parties.

Part 2 reflects the different mechanisms that are associated with the two kinds of lending. Responsibility for fulfilling the requirements of the standard fall on corporate borrowers when "specialized" loans are made to serve their needs. In the case of "standardized" loans, however, the loan originator—typically a bank—bears responsibility for ensuring that the requirements of Part 2 are met.

6.10 How Do Green Loans Work?

Banks and financial institutions originate loans whose proceeds often are used for a defined purpose.

Real estate loans used for the construction or purchase of a property are backed with a double guarantee: mortgage insurance and the mortgage itself. Normally, mortgage loans are low risk unless borrowers lose the source of income which is the basis of repayment. Mortgages that allow borrowers the right to sell their properties reduce that risk, unless the economy in general is depressed and the borrowers' loss of income is matched by a collapse in the resale market. In that case, the property owner or the bank may suffer a loss on the resale of the property.

A "standardized loan" under Part 2 may finance the purchase of specified assets, such as electric vehicles, where the lender works with a manufacturer to extend credit to vehicle buyers, or with home improvement contractors offering weatherization, where the lender's standard terms apply. The distinguishing feature of a standardized green loan is that the lender, rather than the borrower, assumes responsibility for ensuring that the requirements of Part 2 are met.

6.11   ISO's Taxonomy of Eligible Investment Categories

The taxonomy that is detailed in ISO 14030 Part 3 (ISO 2021c)[7] plays a central role in encouraging the financing of the low-carbon economy by providing clarity to issuers and investors about what constitutes a green bond or loan. The taxonomy reduces the risk to issuers, borrowers, and lenders that the market will challenge the environmental benefits of proposed investments. Derisking investment decisions helps protect against reputational harm and accusations of "greenwashing." By using the ISO taxonomy, investors and the institutions who work with them have the strength of an international consensus backing their decision to categorize projects and assets as eligible for financing by green bonds or green loans.

The taxonomy addresses many different sectors of the economy—agriculture and forestry, industry and manufacturing, energy production, transport, and buildings. All economic sectors are relevant to the goal of achieving net-zero emissions by 2050. The taxonomy achieves this by identifying activities, projects, and assets at the subcategory level that can deliver significant environmental benefits. The taxonomy includes relevant criteria and thresholds as well as any categoric exclusions.

ISO recognizes that alternate taxonomies exist around the world and provides the issuers or originators of green debt instruments the option to follow Part 1 or Part 2 while adopting alternate green investment classification systems in the place of Part 3.

6.11.1   Regulatory and Voluntary Taxonomies

Several countries and regions have created taxonomies to support of their climate or energy regulations. The first, called the "Green Bond Endorsed Project Catalogue," was published by the People's National Bank of China in 2015 to align investments according to Chinese environmental objectives (Syn Tao Green Finance 2015). It attracted controversy due to its inclusion of clean coal technologies as eligible projects. In 2020, when this taxonomy was updated, "clean coal" was dropped as an eligible category (Climate Bonds Initiative 2020).

Another prominent taxonomy was created for the European Union. The "Technical annex to the TEG final report on the EU taxonomy" published in March 2020 focused on climate change mitigation and adaptation. It defined criteria to be used for evaluating sustainable financing activities on the one hand. On the other hand, it can be used by companies required to make nonfinancial disclosures under EU law. It also proposed a European label for green investment funds. The taxonomy supports the EU's environmental policy known as the "European Green Deal" (European Council 2021).

---

[7] As of 2021, this document is published as a 2nd "Draft International Standard" with an anticipated final publication date in 2022.

The EU final report of the Technical Expert Group (63 pages) must be read with its technical annex (593 pages) and the EU's taxonomy regulation. The technical annex defines rules for eight major economic sectors.

In Japan, the Ministry of Environment produced "Green Bond Guidelines" in 2017 largely based on the Green Bond Principles. These were updated in 2020 (Japan Ministry of Environment 2020). Canada in 2018 initiated the development of a national taxonomy (ESG Global Advisors 2020), and other countries may also develop taxonomies of their own.

On the voluntary side, the Climate Bonds Initiative (CBI) began in 2013 to develop taxonomies to support the labeling of climate bonds. By 2021, CBI had developed taxonomies to cover 15 defined sectors (Climate Bonds Initiative 2021). Bonds that meet the criteria can be certified by the Climate Bonds Standards Secretariat after successful third-party verification of claims.

ISO 14030 Part 3 was developed in accordance with an international consensus process and is expected to be published in 2022.

Questions for Readers

1. Is a blockchain cryptocurrency a money?
2. Is the current ratio of investment from public and private sources scalable to the economy-wide decarbonization needs of the years up to 2050?
3. Do all types of actors in the finance sector have equal roles to play in greening world economies?
4. How does the ISO 14030-3 taxonomy take environmental aspects into account?
5. What should the role be for green finance in the decades leading up to 2050?

References

Anonymous (2016) Chapitre 1 La théorie ricardienne de l'échange international. https://d.20-bal.com/download/ekonomika-2034/2034.doc. p 3. Accessed 7 May 2021

Bekebrede G et al (2018) Districts: the go2zero simulation game. https://www.mdpi.com/2071-1050/10/8/2602. Accessed 25 Feb 2021

Berger A (2021) Comment la guerre des modèles de financement formate la transition écologique, in responsabilite et environnement, annales des mines Paris n 102 April 2021

Blumsack S (2021) What's behind $15,000 electricity bills in Texas? (The conversation). https://www.santafe.edu/news-center/news/blumsack-whats-behind-15000-electricity-bills-texas-conversation. Accessed 2 June 2021

Bloomberg (2019) Deutsche bank weighs bad-bank unit in revamp: FT bloomberg surveillance TV shows 17 June 2019, 12:04 PM GMT+0200. https://www.bloomberg.com/news/videos/2019-06-17/deutsche-bank-weighs-bad-bank-unit-in-revamp-ft-video. Accessed 6 June 2021

Boursorama (2020) Le rachat d'actions par l'entreprise: est-ce toujours une bonne affaire pour l'actionnaire? 4 December 2020. https://www.boursorama.com/patrimoine/fiches-pratiques/le-rachat-d-actions-par-l-entreprise-est-ce-toujours-une-bonne-affaire-pour-l-actionnaire-5023e3 68e5ddc18cde9ca81a58ab87ab. Accessed 15 April 2021

Build Green (2019) Monnaie locale principe et intérêts. https://www.build-green.fr/monnaie-loc ale-principe-et-interets/. Accessed 16 April 2021

Chernyakhòvskiy I et al (2016) U.S. laws and regulations for renewable energy grid interconnection. https://www.nrel.gov/docs/fy16osti/66724.pdf. Accessed 15 Feb 2021

Chiu L (2020) A brief history of the renminbi thoughtco, 27 August 2020. https://www.thoughtco.com/brief-history-of-the-renminbi-chinese-yuan-688175. Accessed 27 April 2021

Clerc L (2021) "Prise de conscience du risque climatique et de sa dimension systémique" in responsabilité et environnement. Annales des mines Paris n 102 Avril 2021 p 6–9

Climate Bonds Initiative (2017) Bonds and the climate change. The state of the market/Update 2017 India. https://www.climatebonds.net/resources/reports/bonds-and-climate-change-%E2% 80%93-state-market-2017-india-update. Accessed 2 May 2021

Climate Bonds Initiative (2020) Green bond endorsed projects catalogue (2020 Edition) (Draft for consultation). https://www.climatebonds.net/files/files/China-Green-Bond-Catalogue-2020-Con sultation.pdf. Accessed 8 June 2021

Climate Bonds Initiative (2021) Explaining green bonds. https://www.climatebonds.net/market/exp laining-green-bonds. Accessed 5 Feb 2021

Climate Policy Initiative (2018). http://climatepolicyinitiative.org/wp-content/uploads/2018/06/ Publication.png. Accessed 31 Nov 2020

Commission économique pour l'Europe 19 janvier 2016. ECE/MP.EIA/IC/2015/4. https://unece.org/fileadmin/DAM/env/documents/2019/ece/Restart/Belarus/sessions/34/14_oct_report_34_ IC_FR_Report_ece_mp.eia_ic_2015_4_f.pdf Accessed 2 June 2021

Copena D et al (2019) Local economic impact of wind energy development: analysis of the regula-tory framework, taxation, and income for Galician municipalities. https://www.mdpi.com/2071-1050/11/8/2403. Accessed 4 Feb 2021

Crypto-France (2018) Offre de bitcoins: 80% des 21 millions de BTCs ont été émis. 15 January 2018. https://www.crypto-france.com/offre-bitcoins-80-pourcents-21-millions/. Accessed 12 June 2021

Dejean F (2021) Comptabilité et environnement: compter autrement. In: Responsabilité et environnement, annales des mines Paris n°102 April 2021 pp 69–71

Dualsun (2021) The performance and production of a solar panel. https://news.dualsun.com/solar-technology/performance-production-solar-panel. Accessed 22 March 2021

EIA (2021). Independent statistics and analysis. In: U.S. energy information administration (EIA). Texas. State energy profile overview. https://www.eia.gov/state/?sid=TX. Accessed 9 Jun 2021

Ekeland I (2021) La finance à l'heure des limites planétaires, in Responsabilité et environnement, annales des mines Paris. n°102 Avril 2021 p 4

ESG Global Advisors (2020) https://www.esgglobaladvisors.com/post/sustainable-finance-taxono mies-a-spotlight-on-transition. Accessed 5 May 2021

European Central Bank (2012) Virtual currency schemes October 2012. https://www.ecb.europa.eu/pub/pdf/other/virtualcurrencyschemes201210en.pdf. Accessed 14 June 2021

European Council (2021) Taxonomy for sustainable activities. https://ec.europa.eu/info/business-economy-euro/banking-and-finance/sustainable-finance/eu-taxonomy-sustainable-activities_en. Accessed on 5 May 2021

European Parliament (2019) Virtual money: how much do cryptocurrencies alter the funda-mental functions of money? https://www.europarl.europa.eu/cmsdata/190132/PE%20642.360% 20LSE%20final%20publication-original.pdf. Accessed 5 Dec 2020

Gadrey J (2014) La finance pèse-t-elle 100 fois plus que l'économie réelle? 10 fois plus? Bien moins? Blog in alternatives économiques. https://blogs.alternatives-economiques.fr/gadrey/2014/09/13/ la-finance-pese-t-elle-100-fois-plus-que-l-economie-reelle-10-fois-plus-bien-moins. Accessed 7 May 2021

Geary K (2015) The suffering of others. The human cost of the international finance corporation's lending through financial intermediaries OXFAM. https://www.oxfam.org/en/research/suffering-others. Accessed 2 June 2021

Giraud G (2012) Rendre le monopole de la création monétaire aux banques centrales? http://www.revue-banque.fr/risques-reglementations/chronique/rendre-monopole-creation-monetaire-aux-banques-cen. Accessed 2 June 2021

Giraud G (2013) Les dépôts à la banque des Français ne sont pas garantis. https://www.franceculture.fr/emissions/les-matins/ce-que-va-changer-ou-pas-lunion-bancaire. Accessed 12 June 2021

Giraud G (2021) Les modèles économiques et financiers face à la polycrise écologique in responsabilité et environnement. Annales des mines Paris n°102 Avril 2021 p 10

Grandjean A (2021) Réguler la finance, financer le long terme in Chroniques de l'anthropocène. https://alaingrandjean.fr/nos-combats/reguler-finance-financer-long-terme/. Accessed 7 May 2021

Green Climate Fund (2019) Countries step up ambition: landmark boost to coffers of the world's largest climate fund. https://www.greenclimate.fund/news/countries-step-ambition-landmark-boost-coffers-world-s-largest-climate-fund. Accessed 5 Jan 2021

Hajric V, Ballentine C (2021) Cryptocurrencies. Bitcoin rally stirs BofA alarm on 'enormous' surge in energy use. Bloomberg news 22 March 2021. https://www.bloomberg.com/news/articles/2021-03-22/bitcoin-s-carbon-footprint-conveniently-downplayed-during-rally?utm_source=CP+Daily&utm_campaign=c460e64dcb-CPdaily10052021&utm_medium=email&utm_term=0_a9d8834f72-c460e64dcb-110301553. Accessed 11 May 2021

Hetzel J, Tenière Buchot P, Clavet A (1990) La réparation des dommages catastrophiques. Bruxelles 1990 in Bibliothèque de la faculté de droit de l'université catholique de Louvain p 194

Holter M (2016) Norway's sovereign wealth fund drops 52 companies linked to coal. The independent. https://www.independent.co.uk/news/business/news/norway-s-sovereign-wealth-fund-drops-52-companies-linked-coal-a6987931.html. Accessed 22 March 2021

Hurlbut D (2010) Multistate decision making for renewable energy and transmission: an overview. University of Colorado law review. 81, 677. http://lawreview.colorado.edu/wp-content/uploads/2013/11/8.-Hurlbut-Final_s.pdf. Accessed 2 June 2021

Hurlbut D, Getman D (2015) Greening the grid: implementing renewable energy zones for integrated transmission and generation planning. Webinar, December 1, 2015. https://cleanenergysolutions.org/training/greening-grid-implementing-renewable-energy-zones-integrated-transmission-generation. Accessed 2 June 2021

Hurlbut et al (2016) Renewable energy zones: delivering clean power to meet demand. https://www.nrel.gov/docs/fy16osti/65988.pdf. Accessed 12 June 2021

IEA (2020) SDG 7: data and projections. Energy intensity. October 2020. https://www.iea.org/reports/sdg7-data-and-projections/energy-intensity. Accessed 27 Feb 2021

IIF (2021) Task force on scaling voluntary carbon markets: final report. Institute of international finance. https://www.iif.com/tsvcm. Accessed 12 June 2021

ISO (2021a) ISO 14030-1 green debt instruments—process for green bonds. https://www.iso.org/standard/43254.html. Accessed 8 June 2021

ISO (2021b) ISO 14030-2 green debt instruments—process for green loans. https://www.iso.org/standard/75558.html. Accessed 8 June 2021

ISO (2021c) ISO 14030-3 green debt instruments—taxonomy. https://www.iso.org/standard/75559.html. Accessed 8 June 2021

Japan Ministry of Environment (2020) Green bond guidelines, green loan and sustainability linked loan guidelines. https://www.env.go.jp/policy/guidelines_set_version_with%20cover.pdf. Accessed 6 May 2021

Jones L (2020) $1Trillion mark reached in global cumulative green issuance: climate bonds data intelligence reports: latest figures. https://www.climatebonds.net/2020/12/1trillion-mark-reached-global-cumulative-green-issuance-climate-bonds-data-intelligence. Accessed 3 Jan 2021

Kelas K (2021) What is cryptocurrency? Aka crypto. https://cerdasberilmu135.blogspot.com/2021/06/what-is-cryptocurrency-aka-crypto.html. Accessed 8 June 2021

Kusimba C (2017) When—and why—did people first start using money? https://theconversation. com/when-and-why-did-people-first-start-using-money-78887. Accessed 5 Feb 2021

Leiserson G (2017) Issue brief: what is the federal business-level tax on capital in the United States? In: Washington center for equitable growth. https://equitablegrowth.org/what-is-the-federal-bus iness-level-tax-on-capital-in-the-united-states/. Accessed 12 June 2021

Macquarie R et al (2019) Updated view on the global landscape of climate finance 2019. https:// www.climatepolicyinitiative.org/publication/updated-view-on-the-global-landscape-of-climate-finance-2019/. Accessed 31 Nov 2020

Michaels D (2021) Climate change emerges as a compliance focus for SEC updated March 3 2021. The wall street journal. https://www.wsj.com/articles/climate-change-emerges-as-a-compliance-focus-for-sec-11614802615. Accessed 21 March 2021

Millot L (2006) Désormais convertible, le rouble vaut bien plus qu'un kopek in Libération. https:// www.liberation.fr/futurs/2006/07/03/desormais-convertible-le-rouble-vaut-bien-plus-qu-un-kop eck_44882/?redirected=1. Accessed 10 May 2021

Mullin K (2017) Time the green bond market grew up—comment, environmental finance 5 June 2017

Narayanan A (2015) "Private blockchain" is just a confusing name for a shared database. https:// freedom-to-tinker.com/2015/09/18/private-blockchain-is-just-a-confusing-name-for-a-shared-database/. Accessed 8 June 2021

OECD (2016) Pension funds in figures. https://www.oecd.org/daf/fin/private-pensions/Pension-funds-pre-data-2016.pdf. Accessed 10 Dec 2020

OECD (2017) Development aid rises again in 2016 but flows to poorest coun-tries dip. https://www.oecd.org/dac/development-aid-rises-again-in-2016-but-flows-to-poorest-countries-dip.htm. Accessed 5 Jan 2021

Parance B (2021) Les risques juridiques et réputationnels. In: Responsabilité et environnement. Annales des mines Paris n°102 Avril 2021

Pattison D (2015) Social Security cash flows and reserves. Soc Secur Bull 75(1). https://www.ssa. gov/policy/docs/ssb/v75n1/v75n1p1.html. Accessed on 10 May 2021

Piketty T (2021) Le debat eco : cryptomonnaies comment ça marche, bulle spéculative ou vrai changement économique ? France inter 16 avril 2021. https://www.youtube.com/results? search_query=d%C3%A9bat+economique+france+inter+piketty+16+avril+2021. Accessed 16 April 2021

Raval A (2021) Shell's defeat spells trouble for other polluters. In: Financial times, 28 May 2021, p 9

Slattery M et al (2012) The predominance of economic development in the support for large-scale wind farms in the U.S. great plains. https://ideas.repec.org/a/eee/rensus/v16y2012i6p3690-3701. html. Accessed 5 Dec 2020

Snégaroff T (2015) Radio France "2001: une très grande entreprise fait faillite après la révélation de ses mensonges..." https://www.francetvinfo.fr/replay-radio/histoires-d-info/2001-une-tres-gra nde-entreprise-fait-faillite-apres-la-revelation-de-ses-mensonges_1787811.html. Accessed 15 April 2021

Söderberg G. (2018) "Are Bitcoin and other crypto-assets money?" Economic Commentaries. No. 5/2018. 14 March. Sveriges Riksbank, Stockholm.

Syn Tao Green Finance, translators (2015) Notice of the people's bank of China on green finan-cial bonds (PBoC Document No. 39 [2015]). http://www.syntaogf.com/Menu_Page_EN.asp?ID= 21&Page_ID=150. Accessed 8 June 2021

UNFCCC (2021) Race to zero campaign. https://unfccc.int/climate-action/race-to-zero-campaign. Accessed 2 May 2021

US EPA (2021) Green power partnership. Retail electricity markets. https://www.epa.gov/greenp ower/us-electricity-grid-markets. Accessed 8 June 2021

World Bank (2021) Financial sector. https://www.worldbank.org/en/topic/financialsector/overview. Accessed 15 June 2020

Chapter 7
Adaptation

7.1 The International Context for Adaptation

Chapter 1 explains the science behind climate change and describes impacts on weather that are already occurring. One question for organizations and societies at all levels is how to adjust to a changed or changing climate and to make both our built environment and our societies more resilient. The processes employed to respond to the effects of climate change are called "adaptation." Adaptation is frequently paired with the term "mitigation" as both responses to climate change are needed. The former focuses on planning actions to counteract negative impacts from climate change. The latter works to slow and ultimately stabilize climate change at a level that will forestall the most catastrophic impacts that unchecked warming could provoke if efforts to reduce carbon dioxide (CO_2) emissions are not successful.

The 196 parties to the Paris Agreement addressed adaptation in Article 7. The parties "established the global goal on adaptation of enhancing adaptive capacity, strengthening resilience, and reducing vulnerability to climate change, with a view to contributing to sustainable development and ensuring an adequate adaptation response in the context of the temperature goal referred to in Article 2." Article 2 set forth the goal of "holding the global average temperature increase to 2 °C above pre-industrial levels and to pursue efforts to limit the temperature increase to 1.5 °C above pre-industrial levels" (UNFCCC 2015 pp. 22, 25).

While mitigation requires a coordinated global response, given the mixing of emissions of greenhouse gases to an atmosphere shared by countries the world over, adaptation takes place within countries and often at the community or organizational level of societies. Individuals, communities, and countries have long rebuilt infrastructure and economies after natural disasters. What distinguishes these precedents from adaptation is the conscious effort to take the factor of climate change into account in the planning and implementation stages, thereby improving the ability of the object of adaptation to recover from impacts caused by climate-induced effects.

Figure 7.1 presents an adaptation process suitable for use by parties to the Paris

J. C. Shideler and J. Hetzel, *Introduction to Climate Change Management*,
Springer Climate, https://doi.org/10.1007/978-3-030-87918-1_7

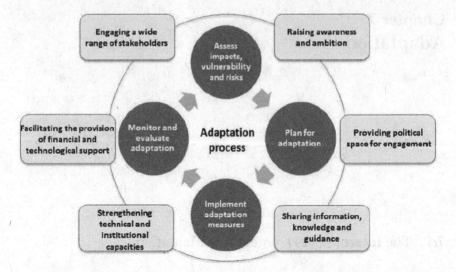

Fig. 7.1 Recommended process for Paris agreement parties addressing adaptation (UNFCCC 2021)

Agreement prepared by the secretariat of the United Nations Framework Convention on Climate Change (UNFCCC).

The UN process is clearly designed for governments as parties to the UNFCCC. It therefore presents a high-level view of the adaptation process. It is less useful for organizations or communities that must engage in planning adaptation measures on a more micro-scale.

The International Organization for Standardization (ISO) has published two documents useful for organizations of all types that are planning adaptation efforts:

– ISO 14090, Adaptation to climate change—Principles, requirements, and guidelines (ISO 2019), and
– ISO 14091, Adaptation to climate change—Guidelines on vulnerability, impacts, and risk assessment (ISO 2021).

Together these documents provide a clear process and guidance to users on how to perform vulnerability and risk assessments, evaluate adaptive capacity, and undertake adaptation actions from a more "bottom-up" perspective.

7.2 Disaster Risks are Related to Climate Change

The IPCC Special Report on the Management of Risks of Extreme Events and Disasters highlighted the need for economies to pursue "climate-resilient development pathways" that contribute to sustainable development (IPCC 2012, p. 52).

Adaptation complements mitigation and takes place in specific geographic contexts that may be international, national, or local.

The IPCC concluded that risk management must consider not only climate factors but also non-climatic environmental and human factors such as those that contributed to a Himalayan disaster that occurred early in 2021 (Science World 2021). In the Himalayan case, environmentalists warned India's Supreme Court in 2014 that the impact study of a hydroelectric dam needed to be revisited, as the inevitable melting of ice in Himalayan glaciers would release water that could sweep through the narrow valleys that were to host the dam. Seven years later, in February 2021, a deluge of rain caused a flood where waters loaded with waste and debris from soil erosion crashed into the dam still under construction, killing 31 people and leaving 165 people missing (Ghosal 2021).

In its Assessment Report 5 (AR5), the IPCC advanced a risk assessment methodology that evaluated hazards, vulnerability, and exposure as the key variables (ISO 2021, pp. 18–19). This updated a vulnerability-based approach that the IPCC had described in its Assessment Report 4 (AR4) in 2007. Adaptation literature may refer to either approach.

The approach described in AR5 characterizes risk as a function of hazards, exposures, and vulnerability, where vulnerability is a function of sensitivity to climate-related stimuli and adaptive capacity. This approach builds upon two classic notions of industrial risk analysis based on severity and probability of occurrence ($R = S \times O$), where R = risk, S = severity, and O = occurrence.

Hazards are "potential sources of harm" (ISO 2021, p. 2). In the Himalayan example, hazards included the potential for melting snow and heavy rains, erosion-prone riverbanks, and the accumulation of debris.

Exposure indicated the "presence of people, livelihoods, species or ecosystems, environmental functions, services, resources, infrastructure, or economic, social, or cultural assets in places and settings that could be affected" (ISO 2021, p. 3). In the Himalayan example, the dam under construction and the people gathered to build it were exposed to the hazards.

Vulnerability means the "propensity or predisposition to be adversely affected" (ISO 2021, p. 3). It considers the sensitivity of systems or species. In the case of the Himalayan disaster, vulnerability stemmed from the narrowness and steepness of the valley which would accelerate the velocity of floodwaters placing both dam works and people at great risk.

Climate change most likely played a role in the Himalayan disaster. As foreseen by project critics, warmer temperatures accelerated the melting of snow and ice and likely intensified precipitation events. Impacts from the flooding event included the loss of life and the destruction of the dam and other downstream structures. The force of the raging water liquefied the riverbanks.

A diagram presented as Fig. 7.2 expands upon the risk assessment concept first advanced in AR5. It shows the relationship between risks and impacts. It recognizes contributing factors that were described in the AR5 diagram as "climate" and "development." The latter category, shown on the right-hand side of the figure, was updated in the ISO 14091 document to "socioeconomic processes."

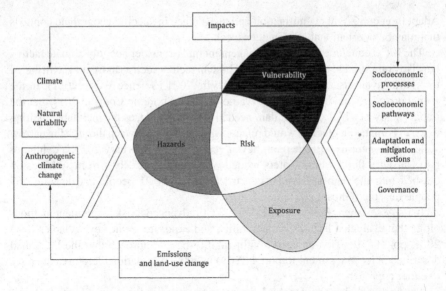

Fig. 7.2 Intersection of factors that lead to disaster risk (ISO 2021, Fig. A.3, adapted from AR5).
Figure ©ISO. This material is reproduced from ISO 14091:2021, with permission of the American
National Standards Institute (ANSI) on behalf of the International Organization for Standardization.
All rights reserved

ISO 14091 subdivides "socioeconomic processes"[1] into socioeconomic pathways,
adaptation and mitigation actions, and governance. One description of socioeconomic
processes (Anonymous no date) includes the following three elements:

– Population dynamics and demography
– Human–environment interactions
– Urbanization, globalization, and ecological footprint.

With respect to the Himalayan disaster, more construction workers and adjacent
villagers were exposed to risk than architects and engineers, project managers, planners, regulators, and financiers. The former were disproportionately represented in
the casualties from the event.

The construction of the dam could be viewed as a "mitigation action" if the intent
of building the dam was to replace fossil-fueled energy sources with low-carbon
hydro.

Critics of the dam project cited governance concerns. "Uma Bharti, a former
minister for water resources and river development in the government of Prime
Minister Narendra Modi, said she had warned against placing a hydroelectric project
on the river so close to the Himalayas." She had previously warned that tributaries
to the Ganges River in the Himalayas were in sensitive areas and inappropriate for

---

[1] The term "socioeconomic"—"of, relating to, or involving a combination of social and economic
factors", dating to 1883 in American usage (Websters 1988)—should not be confused with the
"research tradition" of socioeconomics described by Hellmich (2017).

hydro development. Environmentalist Anil Joshi attributed the triggering event to a glacier avalanche which was due to climate change. He faulted the government's decision to build the dam so close to the glacier and noted that "now this water is flowing with cyclonic speed" (Mashal and Kumar 2021).

The geological phenomenon of liquefaction, which played a role in the Himalayan disaster, is worthy of further elaboration considering the climate-induced intensification of precipitation events. Liquefaction is observed in many environments such as coastal areas, riverbanks, but also, increasingly, in mountainous areas. This phenomenon is the result of the pressure of the oceans on shorelines and of rivers on valley walls, especially when mountainous areas have been developed to manage rivers for flood control or power generation. Ocean waves are powerful forces when added sands are projected violently on shores. In the case of rivers, water volumes swollen by heavier precipitation events create accelerated flows that multiply their force and potential for damage. In many areas, the acceleration of water flows is due to the increased impermeability of soils that results from the extension of paved and built environments.

Figure 7.3 illustrates the liquefaction phenomenon. Liquefaction can attack the soil supporting buildings or other structures leading to the collapse of structures and the washing away of soils.

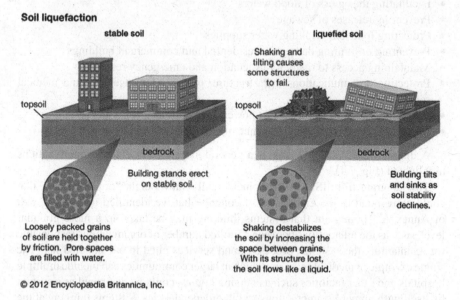

© 2012 Encyclopædia Britannica, Inc.

Fig. 7.3 Explanation of liquefaction and its impacts (Encyclopedia Britannica 2021)

A primary purpose of adaptation is to make the built environment more resilient, and responses to climate-induced impacts more effective. The concept is not new. Humans practicing agriculture have modified the natural environment for centuries if not millennia. To reduce illness associated with marshes, Dutch engineers drained and managed "polders." These low-lying areas were made suitable for agriculture by surrounding parcels of land with dikes and using pumps to move excess water into canals. In this way, the Dutch, the Venetians, and others have erected protections against the intrusion of the sea since the twelfth century (Van Schoubroeck and Kool 2010).

7.3 Systems Thinking

Systems thinking is introduced in ISO 14090 as a principle: "Climate change adaptation processes include an understanding of cross-cutting (systemic) issues of the organization by examining internal and external interdependencies and linkages, for example through cause-and-effect relationships" (ISO 2019, p. 5).

Sea-level rise is an example of a slow-onset impact of climate change, for example, on the US East and Gulf coasts, where cities like Miami, Florida, and Virginia Beach, Virginia, already face regular flooding. Systems thinking provides a tool for thinking beyond a seawall or raised structure. Planners in such cases should consider a variety of cross-cutting issues, such as:

- Facilitating the egress of flood waters
- Preventing releases of sewage
- Protecting food and drinking water supplies
- Preventing or limiting damage to residential and commercial buildings
- Maintaining access to roads for evacuation and emergency response
- Protecting communications infrastructure, public safety agencies and medical services
- Hardening the power grid against power interruptions
- Installation, testing, and maintenance of emergency power generators.

A diagram in ISO 14090 depicts a general systems concept with interventions highlighted (Fig. 7.4).

This diagram from ISO 14090 coincidentally shows eight "organizations" that could correspond to the eight areas of concern that we identified above. The text of Annex A.4 points out that systems thinking may be taken to a more granular level, such as the interactions between a limited number of organizations responsible for maintaining the systems, processes, and services cited in our bulleted list. The single example of medical services, which in larger communities can include multiple hospitals and other facilities such as nursing homes, could be described as a system by itself with numerous interactions and interdependencies. Systems thinking at the community level may therefore spawn more granular systems thinking at the sector or organizational levels.

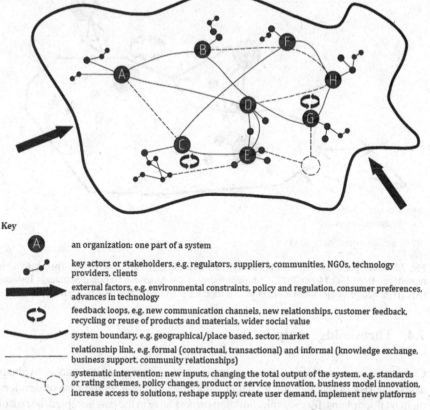

Key

(A)	an organization: one part of a system
	key actors or stakeholders, e.g. regulators, suppliers, communities, NGOs, technology providers, clients
	external factors, e.g. environmental constraints, policy and regulation, consumer preferences, advances in technology
	feedback loops, e.g. new communication channels, new relationships, customer feedback, recycling or reuse of products and materials, wider social value
	system boundary, e.g. geographical/place based, sector, market
	relationship link, e.g. formal (contractual, transactional) and informal (knowledge exchange, business support, community relationships)
	systematic intervention: new inputs, changing the total output of the system, e.g. standards or rating schemes, policy changes, product or service innovation, business model innovation, increase access to solutions, reshape supply, create user demand, implement new platforms

Fig. 7.4 Systems concept (ISO 2019, Fig. A.1). Figure ©ISO. This material is reproduced from ISO 14090:2019, with permission of the American National Standards Institute (ANSI) on behalf of the International Organization for Standardization. All rights reserved

ISO 14090 illustrates this concept by depicting a smaller subsystem boundary that encompasses only two organizations which are in turn strongly influenced by two organizations that lie outside the subsystem boundary (Fig. 7.5).

ISO 14090 (2019) illustrates how elements identified by systems mapping of a subsystem might relate and interact:

- Organization E might be an energy transmission and distribution company.
- Actor z could be a back-up power supply connection.
- Organization G might be a local solar energy facility.
- Organization D might provide interconnected energy from outside the local energy supply grid.
- Intervention Y might involve sourcing energy from D to replace energy generation sources shut down due to temporary inundations in areas of the transmission and distribution system not affected by floodwaters.

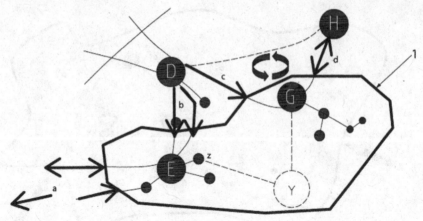

Fig. 7.5 Example of a subsystem boundary (ISO 2019, Fig. A.2). Figure ©ISO. This material is reproduced from ISO 14090:2019, with permission of the American National Standards Institute (ANSI) on behalf of the International Organization for Standardization. All rights reserved

7.4 Thresholds Analysis

Climate change is affecting the way organizations, communities, and individuals assess the risk of maintaining a system without changing it. In the case of development on floodplains, for example, building owners have in the past accepted the risk of occasional flooding whose damages could be mitigated through the purchase of flood insurance. If building owners located in a floodplain experienced a flood, flood insurance provided money to rebuild. In some designated floodplains, purchase of flood insurance is mandatory. As climate change has increased the frequency of flooding and economic damages have ballooned, the economic case for maintaining the status quo in areas at risk has become more complicated. A "threshold" for changing the paradigm has been approached or exceeded, causing changes in behaviors on the part of economic actors. Changes can include rezoning to prohibit reconstruction or to impose new conditions on rebuilding such as the elevation of structures on stilts.

Annex B of ISO 14090 (2019) describes thresholds analysis. It identifies five steps that can lead either to incremental or transformative changes. The five steps include:

- characterize the system
- research possible climate changes
- identify thresholds
- assess resilience
- identify suitable indicators.

ISO 14090 prefers the term "thresholds analysis" to "tipping point," a term frequently found in adaptation literature that has the same meaning (ISO 2019, p. 24).

In the following paragraphs, we illustrate the point that maintaining the status quo in the face of a changing climate was no longer acceptable. Thresholds had been reached.

7.5 Construction Along Coastal Areas and Rivers

The latest research suggests that by 2100, as much as 60% of oceanfront communities on the East and Gulf coasts of the USA may experience chronic flooding from climate change (Bennington-Castro 2017). Seawalls will provide limited protection due to erosion and liquefaction (see box). Seawalls can provide some protection from hurricanes to properties located behind them, but their extent typically is limited as sections of shoreline may remain exposed. In some cases, seawalls lead to coastal erosion, creating a vector for damage rather than providing the intended shoreline protection (Balaji 2017).

Seawalls may be built with bulkheads. These are less efficient against waves, but they can be integrated in construction designs, including for floating and amphibious housing. Shorelines can incorporate mangroves which reintroduce natural plant and animal species to help protect the terrain.

Floating and amphibious houses are built to be situated in a water body and are designed to adapt to rising and falling water levels. Floating houses are moored permanently in the water, while amphibious houses are situated above normal high-water lines and are designed to float when the water levels are abnormally high. Amphibious homes are usually fastened to flexible mooring posts and rest on concrete foundations. If the water level rises, they can move upward and float. The fastenings to the mooring posts limit the motion caused by the water. These types of houses are popular in highly populated areas where there is a high demand for houses near or in water. Because floating or amphibious houses adapt to rising water levels, they are very effective in dealing with floods.

Living on water can also reduce the negative effects of heat and may improve the quality of life of residents. Floating houses have already been built in various countries, like the Netherlands and the United Kingdom, and amphibious houses in the Netherlands. The scale of this type of development can vary from an individual house to large groups of dwellings to, theoretically, full-blown floating cities. Options of this kind have been tried mainly in inland waters, but marine applications are also possible (DG CLIMA project 2016) (Fig. 7.6).

Located on the Thames River, the UK's first amphibious house was constructed in 2014. Baca Architects designed the home for a couple who wanted to live on a flood-prone island in the river, integrating a terraced landscape that acts as an early warning system that the waters are rising. In Louisiana, the Buoyant Foundation Project devised a solution to retrofit existing homes in post-Katrina New Orleans in anticipation of future storms and floods, allowing the structures to lift off the ground

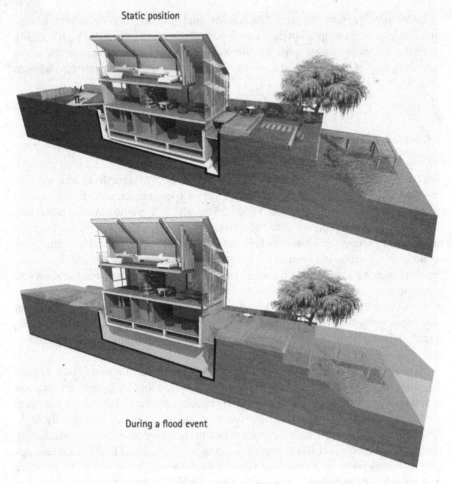

Static position

During a flood event

Fig. 7.6 Amphibious housing in floating area (Rogers 2014)

in an emergency. Buoyancy blocks would be installed beneath the subframe of the home, while four corner guideposts keep the building in place as it rises with the water (Rogers 2014).

Due to the need for arable soil to produce food for humans, land use regulations can be adapted to limit surface areas available for construction. Engineers, architects, and environmentalists can collaborate to build adapted cities. People manage their social-ecological systems according to their often-limited perceptions of the opportunities and risks, and how they value the alternatives (Elmqvist et al. 2014).

After each disaster such as Hurricane Katrina or Superstorm Sandy, promoters, builders, architects, and urbanists propose new concepts for building safer but some-times encounter obstacles. "Our switchgear is in the basement and I do not know how one can move that elsewhere," said Nathan Berman, the principal of Metro Loft

Management, the building landlord. "It is a matter of economics—I am not sure too many developers would want to compromise lucrative space elsewhere in the building for a storm that was hopefully just a 25- or 50-year event" (Satow 2013).

As always, short-term gains are preferred due to the lack of risk analysis, but scientific risk analysis often has been performed and can be referred to, as for example, in "The uncertain future of coasts," a chapter in World Ocean Review (2010) which observed that 8% of the US population resides in low-lying coastal regions (WOR1 2010, p. 73). The WWF report "Mega-Stress for Mega Cities, A Climate Vulnerability Ranking of Major Coastal Cities in Asia" (WWF 2009) serves as another reference document.

Modern engineering techniques expand possibilities for building in challenging environments. The building "Foundation Louis Vuitton" in Paris, France, designed by Franck O. Gehry resembles a sailing ship. It houses a modern art museum that keeps the Bernard Arnaud modern art collection in a safe concrete structure built to resist fire, flooding, and other risks.

The scale model of the building (see Fig. 7.7) shows the location of critical service equipment on the building's upper level. This ensures full protection of the building and its collections during adverse weather events. It is an example of adapting to known and expected risks. One open question, of course, is when will threshold analysis dictate that similar protective measures need to be extended to nearby existing residential and commercial buildings.

Fig. 7.7 Numerical scale model of Foundation Louis Vuitton Paris (*Source* Digital Project BIM Model of the Foundation Louis Vuitton, Courtesy of STUDIOS Architecture, Local Architect for Gehry Partners)

As we argue later in this chapter, building regulations in at-risk areas should be modified to extend similar protections to all classes of people and not just to the favored few.

Human ingenuity does not always rely on modern technology. In Martinique (West Indies), people who live close to the ocean have learned to leave their homes and keep all their doors open to let hurricane floodwaters enter and recede. They paint a number on their tin roofs which may be displaced by hurricane-force winds. Homeowners can be identified by the number and will pay after the end of the storm to have their roofs returned to them (Hetzel 2004).

7.6 Climate Change Scenarios and Pathways

The IPCC AR5 focus on "climate-resilient development pathways" promotes the use of climate scenarios with projections that scientists rate according to their confidence in them (high, medium, and low levels of confidence). This approach acknowledges that uncertainty relating to climate change impacts and vulnerabilities exists for the reasons described in ISO 14091 (ISO 2021, p. 20):

- climate change will vary in magnitude depending upon future levels of greenhouse gas emissions, which are unknown
- climate extremes are more difficult to predict than slow-onset and long-term trends
- models used for impact assessments reflect uncertainties
- estimating future adaptive capacities is challenging.

The pathways described by AR5 have multiple dimensions. In addition to addressing risks associated with climate change adaptation and mitigation, AR5 recommends that selected pathways should aim to meet other goals of sustainable development, including poverty eradication and reducing inequalities. AR5 acknowledges that any feasible pathway that remains within the 1.5 °C temperature-rise band involves synergies and trade-offs (IPCC 2012, p. 52). Pathways, according to AR5, should be broadly defined as iterative processes for managing change within complex systems in order to reduce disruptions and enhance opportunities associated with climate change. They reflect "a series of adaptation choices involving trade-offs between short-term and long-term goals and values" (IPCC 2012, p. 64).

7.7 Impact Chains

Impact chains are a tool to help organizations evaluate hazards, exposure, and vulnerability related to climate change. They can also be used more generally for analyzing risk at critical points of procurement, operations, and sale. Many industrial companies located in Europe and North America realized too late how vulnerable their supply chains were to disruption after a tsunami hit Japan in 2011 and disrupted the country's power grid. General Motors in the USA had to close manufacturing plants because Japanese suppliers were shut down for days as a result of rolling blackouts and considerable disruption to the transportation systems (Baxter 2016). The impact from the climate-related event was exacerbated by "just-in-time" delivery schedules that reduced parts inventories and magnified the supply impacts of production disruptions. Flooding in Thailand in 2011 also interrupted supply chain deliveries of electronic components and had cascading impacts on computer manufacturers and others.

The guidance in ISO 14091 (2021) recommends that planners evaluate how climate change may influence hazards, exposure, and vulnerability, and lead to impacts. Sensitivity is considered as well as the capacity to adapt and make a system more resilient. Figure 7.8 illustrates the concept with reference to the excess precipitation that caused flooding in Thailand in 2011 (Fig. 7.8).

The flooding event in Thailand in 2011 was unparalleled in its magnitude and even the best adaptation planning and investments in resilience may not have been sufficient to avoid disruption of manufacturing production. This is because flooding impacted the homes and mobility of workers, transportation modes, the electricity grid, and virtually every other resource needed to maintain manufacturing operations and normal daily life in the affected areas. Consequently, impact chains can become very complex when all the nodes in a system are separately evaluated and their interrelationships analyzed. Haraguchi and Lall (2015, pp. 257–258) describe the flooding and a number of confounding factors that intensified its impacts, as well

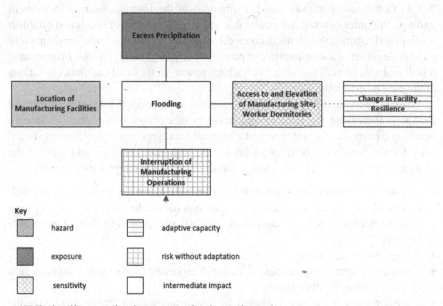

NOTE: The dotted line means that adaptation actions have been implemented.

Fig. 7.8 Example of an impact chain (adapted from ©ISO 2021, Fig. C.2)

as the disruptions the event caused to manufacturers reliant on Thai production part suppliers.

To reduce costs, many manufacturers in developed countries have outsourced the production of parts and components, particularly in the first two decades of the twentieth century. A reevaluation of that strategy has gained attention in the third decade of the twenty-first century as companies reassess the costs of supply chain disruption, political risk associated with operating in some countries and economies, and security risks, including cyber security. Even without the factor of climate change, manufacturers are using risk assessment to address such questions as access to raw materials, water resources, geographic diversity of suppliers, and dependence upon proprietary technologies.

Climate change is an additional risk that also has multiple dimensions. Direct effects include extreme weather events such as those experienced during Thailand's flooding in 2011. Carbon emissions present another risk as countries and regions consider adopting carbon border adjustment taxes. This policy mechanism (see Chap. 2) is intended to protect a country's domestic suppliers from unfair competition that arises when imported goods are produced in countries that do not tax carbon.

7.8   Creating Redundant Infrastructure

The architecture of the Internet was designed with redundancy in mind. Companies that do thousands of transactions per minute over the Internet, such as investment trading companies or retail companies that sell large quantities of goods and establish resilient and redundant systems to connect to the Internet. Such companies do not rely on a single server or a single source of electricity to power their systems. Instead, they install multiple servers and invest in back-up power, so they can maintain operations in the case of grid failure.

Planned redundancy is one way to improve resilience. We know that climate change is increasing the number and intensity of extreme events such as storms, floods, and hurricanes. For this reason, planners, managers, and policy makers should consider the long-term benefits of hardening important systems and against the impacts resulting from climate change risks. Hardening strategies may include:

- Residential, commercial, and industrial buildings are designed (or retrofitted) to ensure key building systems such as electricity, heating and cooling, and telecommunications and Internet are robust and protected from foreseeable hazards.
- Energy and water supply sources are diversified.
- Waste and wastewater treatment is managed to prevent downstream contamination during extreme weather events.
- Transportation routes are designed to remain accessible during most extreme weather events.
- Food supply sources are diversified and supply channels strengthened.

- Critical medical and pharmaceutical products are stockpiled and safeguarded.

Human deployment of science and technology can help protect societies against natural risks and support sustainable development. Financial investments (see Chap. 6) can be directed toward green projects and assets which can aid the transition to a low-carbon economy. Greening urban environments where more than 50% of the world's populations live is necessary to reduce GHG emissions and support climate change mitigation.

7.9   Restoring Environmental Amenities in Urban Centers

A case study in modern urbanization can be found in Korea. The Cheong Gyecheon district of Seoul (Republic of Korea) was divided by a concrete urban highway built in the 1950s. It contributed to traffic congestion and air pollution. City leaders reimagined how this corridor could be improved and chose the bold plan to remove the highway and create an artificial urban river park stretching 5.8 km through a historical part of Seoul. The route restored a former riverbed which by the 1950s had become nothing more than a street sewer. The public works undertaken to remove the highway and rebuild a water course through the district allowed the city to excavate historical sites and preserve cultural knowledge. The climate benefit was to reduce ambient air temperature in the neighborhood by 2 °C (Seoul Solution 2014).

The project also revitalized entertainment venues in the center of Seoul (Back to a Future Seoul 2006). The project, which had strong public support, achieved multiple objectives:

- excavate and preserve cultural artifacts found during historical archeological excavations that are now displayed in a new museum
- maintain neighborhood activities such as street shops and include the local community in the development of the project
- involve the wider community through a "tree planting Sunday" where inhabitants from all parts of Seoul could plant trees along of the river
- create a pleasant park along the artificial river through which moving water remains oxygenated.

Project Before:

See Fig. 7.9.

Project After:

See Fig. 7.10.

This project could only be completed with the support and leadership of the mayor (future President of the Korean Republic). The initial investment of $280 million was the beginning of a major urban planning project that grew to $1.9 billion.

The benefits were (Landscape Performance series 2010):

Fig. 7.9 View of Cheong Gyecheon in Seoul (Republic of Korea) prior to project implementation (In-Keun Lee 2006)

Fig. 7.10 View of Cheong Gyecheon in Seoul (Republic of Korea) after project implementation (In-Keun Lee 2006)

- 100% of the concrete from the highway demolition was recycled in the construction of the riverbed infrastructure.
- Provides flood protection for up to a 200-year flood event and can sustain rises in water levels at rates of up to 118 mm/hour.
- Increased overall biodiversity by 639% between the pre-restoration work in 2003 and the end of 2008 with the number of plant species increasing from 62 to 308, fish species from 4 to 25, bird species from 6 to 36, aquatic invertebrate species from 5 to 53, insect species from 15 to 192, mammals from 2 to 4, and amphibians from 4 to 8.
- Reduced the urban heat island effect with temperatures along the stream 3.3–5.9 °C cooler than on a parallel road 4–7 blocks away. This results from the removal of the paved expressway, the cooling effect of the stream, increased vegetation, reduction in auto trips, and a 2.2–7.8% increase in wind speeds moving through the corridor.
- Reduced small-particle air pollution by 35% from 74 to 48 μg per cubic meter. Before the restoration, residents of the area were more than twice as likely to suffer from respiratory disease as those in other parts of the city.
- Contributed to 15.1% increase in bus ridership and 3.3% in subway ridership in Seoul between 2003 and the end of 2008.
- Attracts an average of 64,000 visitors daily; of those, 1,408 are foreign tourists who contribute up to 2.1 billion won ($1.9 million USD) in visitor spending to the Seoul economy.
- Increased the price of land by 30–50% for properties within 50 m of the restoration project—double the rate of property increases in other areas of Seoul.
- Increased the number of businesses by 3.5% in Cheong Gyecheon area during 2002–2003, which was double the rate of business growth in downtown Seoul; increased the number of working people in the Cheong Gyecheon area by 0.8%, versus a decrease in downtown Seoul of 2.6%.

7.10 Water Management in Venice

The Venice lagoon has faced periodic flooding for more than a century, but climate change has increased that frequency. For example, in the early twentieth century, Venetians could expect flooding to occur on average seven times per year (Van Boom 2020). To hold the Adriatic Sea at bay, project MOSE (the Italian name for the biblical Moses) installed physical barriers to protect the historic city against rising tides.

The barriers are intended to protect the Venice lagoon against abnormally high Adriatic Sea tides. These normally vary from around 1 m at high tide to about 0.15 m at low tide. In recent times, however, some 100-year tides have risen to around 1.8 m. After floods damaged historical churches and palaces, Venetian authorities initiated project MOSE in 2002 at a planned cost of approximately $2 billion. After construction delays and other issues, the final cost subsequently grew to about $8 billion.

The underlying technology and control of the barrier have not fully met expectations. According to design specifications, the MOSE barrier should activate within 30 min and protect against high tides of 3 m. It failed, however, to activate on December 19, 2019, when the tide rose only to 1.3 m, and Venice was once again flooded. The system worked as designed for the first threatened flood of 2020, but then failed to activate two months later during another high-water event (Davidson 2020). Worse, it had taken Italian officials so long to design and build the barriers that critics said they had not been built high enough: "MOSE 'might be able to avoid flooding for the next few decades, but the sea will eventually rise to a level where even continuous closures will not be able to protect the city from flooding. The question is not if this will happen, but only when it will happen'" (Davidson, quoting from a 2011 UNESCO report).

7.11 Forest Preservation

Genetic improvements of tree seeds present a solution against climate change. The UK government in 2019 provided the following advice: "Maintaining genetic variation in our tree species is important; we do not know with certainty the environmental pressures our trees will face, but the greater the variation, the more likely it is that populations are able to survive and even thrive in the new and changing conditions. Natural selection, via natural regeneration, is important and can drive site-based adaptation. If there is extremely limited natural regeneration, for example due to a lack of seed trees or grazing and browsing impacts, then adaptation will not take place. If these issues cannot be addressed, then appropriate tree stock will need to be brought in to help drive site-based adaptation. Planting appropriate southerly provenance is a valid choice for maintaining and enhancing timber production but is not proven for enhancing resilience by increasing the genetic diversity on site, since most tree populations are already genetically diverse. There is some evidence that shows reduced susceptibility to drought but also an increased frost risk from some more southerly choices" (Forestry Commission 2019).

7.12 Adaptive Capacity

An implied theme of this chapter can be summarized in the term "adaptive capacity." This term means the "ability of systems, institutions, humans, and other organisms to adjust to potential damage, to take advantage of opportunities, or to respond to consequences" (ISO 2021, p. 3). Our example of the circumstances that led to the disaster in the Himalayan valley revealed a low adaptive capacity, as voices expressing misgivings about the suitability of the site chosen for the dam project went unheeded.

The leaders of Venice and Seoul displayed greater adaptive capacity as they brought to conclusion projects that buttressed water defenses as they were understood at the time the project was initiated (Venice) or imagined the benefits of restoring a watercourse and associated amenities in an urban environment (Seoul).

The adaptation challenges arising from climate change surpass the normal bounds of "risk management." Adaptation to climate change requires an understanding of dynamic climate processes that extend beyond what we know from past experience. Some impacts, such as intensified weather events, drought, and extreme heat, are manifest now, though they will intensify more with greater global warming. Others fall in the category of "slow-onset" changes, such as sea-level rise. ISO 14091 categorizes levels of complexity as "simple," "complicated," and "complex" (ISO 2021, p. 36), and many of the current adaptation challenges involve multiple interorganizational dependencies and vulnerabilities. Humans can adapt using a variety of strategies, such as modifying the built environment, reconceiving supply chains, and substituting low-carbon or recycled materials. Plants and animals may adapt by migrating to higher elevations or latitudes; coral reefs may die.

ISO 14091 distinguishes levels of adaptive capacity as "medium," "high," and "very high" (ISO 2021, pp. 36–37). These categories correspond to the ability of organizations to manage adaptation to climate change with varying levels of complexity and over different periods of time. The latter are defined as short term (up to 15 years), medium term (up to 30 years), and long term (beyond 30 years). At the different levels of adaptive capacity, organizations are less or more suited to manage impact(s) that are likely to change over the life of an adaptation action. Higher levels of adaptive capacity take into account greater levels of complexity that the organization will face in mitigating climate-related risk (ISO 2021, p. 35). The greater the adaptive capacity, the more likely it is that an organization can manage the interdependencies that may determine whether chosen responses will be effective (ISO 2021 p. 34).

The need for adaptation as well as mitigation is acknowledged in ISO's Guide 84 (2020). "Adaptation to climate change represents adjustments in ecological, social or economic systems in response to actual or expected climate stimuli or their effects or climate change impacts, risks and opportunities, with a subsequent improvement in resilience" (ISO 2020, p. 10). Mitigation of climate change is not enough: "Measures to adapt to the current new climate and the climates in the medium- to long-term future are clearly also required" (ISO 2020, p. 10). Organizations planning to survive in a changing world are called to assess and, if necessary, augment their adaptive capacity, to meet the challenges that climate change presents.

Questions for Readers

1. How are climate change mitigation and adaptation related?
2. Can you identify situations in which liquefaction has impacted natural landscape features or the built environment?

3. To what extent do you believe that your personal residence is adapted to withstand extreme weather events?
4. How do you assess the trade-offs of living near water?
5. How would you rate the adaptive capacity of your community or your employer?

References

Anonymous (no date) Portion of an unidentified syllabus found via search at http://www.columbia. edu/itc/sipa/esspm/life_land/client_edit/Section1_Part2a.html. Accessed 17 Apr 2021

Balaji R, Sathish Kumar S, Misra A (2017) Understanding the effects of seawall construction using a combination of analytical modelling and remote sensing techniques: case study of Fansa, Gujarat, India. Int J Ocean Clim Sci Technol Impacts. https://doi.org/10.1177/175931311771 2180. Accessed 28 Feb 2021

Back to Future Seoul (2006) Cheong Gye Cheon restoration project Seoul Metropolitan Government. In Seoul H (eds) Chong Gye Cheon Museum, Seoul, Republic of Korea

Baxter D (2016) Supply chain disruption: the bad, the ugly, & the future, real-time visibility in supply chain. Supply Chain 247, 18 October 2016. https://www.supplychain247.com/article/sup ply_chain_disruption_the_bad_the_ugly_the_future. Accessed 25 Apr 2021

Bennington-Castro J (2017) Walls won't save our cities from rising seas. Here's what will: 'green' approaches may be the best way to protect coastal communities from flooding associated with climate change. July 27, 2017, 4:43 PM CEST/Updated Sept. 19, 2017, 6:30 PM CEST (eds) CBS news in https://www.nbcnews.com/mach/science/walls-won-t-save-our-cities-rising-seas- here-s-ncna786811. Accessed 28 Feb 2021

Davidson J (2020) 'You're responding to yesterday's disasters': the $6 billion floodgates of Venice may not be enough. Curbed 16 December 2020. https://www.curbed.com/2020/12/venices-usd6- billion-mose-floodgates-may-not-be-enough.html. Accessed 23 Apr 2021

DG CLIMA project (2016) Adaptation strategy of European cities. Published in climate ADAPT Jun 07 2016. Last Modified in Climate ADAPT Mar 04 2020. https://climate-adapt.eea.eur opa.eu/metadata/publications/eu-cities-adapt-adaptation-strategies-for-european-cities-final- report. Accessed 28 Feb 2021

Elmqvist T et al (2014) Chapter 2 history of urbanization and the missing ecology in http://www. robertcostanza.com/wp-content/uploads/2017/02/2013_Elmquist_et-al._CBO_Chapter2.pdf. Accessed 1 Mar 2021

Encyclopedia Britannica (2021) Soil liquefaction. https://www.britannica.com/science/soil-liquef action/images-videos. Accessed 23 Feb 2021

Forestry commission (2019) UK government policy advice https://assets.publishing.service.gov. uk/government/uploads/system/uploads/attachment_data/file/809733/Policy_Advice_Note_F inal180619.pdf. Accessed 30 Mar 2021

Ghosal A (2021) Himalayan glacier disaster highlights climate change risks. https://apnews.com/ article/climate-climate-change-courts-avalanches-india-7be7a76eea4d497b22609ff3d5194e69. Accessed 15 Feb 2021

Hetzel J (2004) Note environnementale et développement durable, pour la restructuration des quartiers Bergevin. Unpublished, Henri IV et Chanzy à Pointe à Pitrela

Haraguchi M and Lall U (2015) Flood risks and impacts: a case study of Thailand's floods in 2011 and research questions for supply chain decision making. Int J Disaster Risk Reduct 14(3):256–272. https://doi.org/10.1016/j.ijdrr.2014.09.005

Hellmich S (2017) What is socioeconomics? an overview of theories, methods, and themes in the field. Forum Social Econ 46(1):3–25. https://doi.org/10.1080/07360932.2014.999696

International Architecture and Design Firm. In: STUDIOS Architecture. https://studios.com/. Accessed 19 June 2021

IPCC (2012) Special report on the management of risks of extreme events and disasters. https://www.ipcc.ch/report/managing-the-risks-of-extreme-events-and-disasters-to-advance-climate-change-adaptation/. Accessed 28 May 2021

ISO (2019) ISO 14090 Adaptation to climate change—principles, requirements and guidelines. https://www.iso.org/standard/68507.html. Accessed on 28 May 2021

ISO (2020) Guide 84 guidelines for addressing climate change in standards. https://www.iso.org/standard/72496.html. Accessed 28 May 2021

ISO (2021) ISO 14091 adaptation to climate change—guidelines on vulnerability, impacts and risk assessment. https://www.iso.org/standard/68508.html. Accessed 28 May 2021

Landscape Performance series (2010) Cheong Gyecheon stream restoration project. https://www.landscapeperformance.org/case-study-briefs/cheonggyecheon-stream-restoration. Accessed 13 March 2021

Lee I-K (2006) Cheong Gye Cheon restoration project: a revolution in Seoul, Seoul Metropolitan Government. https://seoulsolution.kr/ko/content/cheong-gye-cheon-restoration-project. Accessed 14 June 2021

Mashal M, Kumar H (2021) Glacier bursts in India, leaving more than 100 missing in floods. New York Times Feb. 7, 2021 Updated Feb. 8, 2021. https://www.nytimes.com/2021/02/07/world/asia/india-glacier-flood-uttarakhand.html. Accessed 18 April 2021

Rogers S (2014) Amphibious architecture: 12 flood-proof home designs filed under houses & residential in the architecture category in web urbanist. https://weburbanist.com/2014/10/20/amphibious-architecture-12-flood-proof-home-designs/. Accessed 28 Feb 2021

Satow J (2013) The generator is the machine of the moment in New York Times Jan 11. 2013. https://www.nytimes.com/2013/01/13/realestate/post-sandy-the-generator-is-machine-of-the-moment.html. Accessed 1 March 2021

Science World (2021) Liquefaction. https://www.scienceworld.ca/resource/liquefaction/. Accessed 23 February 2021

UNFCCC (2015) Paris Agreement. https://unfccc.int/sites/default/files/english_paris_agreement.pdf. Accessed 28 May 2021

UNFCCC (2021) What do adaptation to climate change and climate resilience mean? https://unfccc.int/topics/adaptation-and-resilience/the-big-picture/what-do-adaptation-to-climate-change-and-climate-resilience-mean. Accessed 16 Apr 2021

Van Boom D (2020) Venice's desperate 50-year battle against floods in https://www.cnet.com/features/venices-desperate-50-year-fight-against-floods/. Accessed 25 Apr 2021

Van Schoubroeck F, Kool H (2010) The remarkable history of polder systems in The Netherlands. http://www.fao.org/fileadmin/templates/giahs/PDF/Dutch-Polder-System_2010.pdf. Accessed 12 Feb 2021

Websters (1988) Webster's ninth new collegiate dictionary. Merriam Webster, Springfield, MA

World Ocean Review 1 (2010) Living with the oceans. The uncertain future of the coasts Published by Maribus in cooperation with The Future Ocean, Kiel Marine Sciences. https://worldoceanreview.com/wp-content/downloads/wor1/WOR1_en.pdf. Accessed 28 Feb 2021

WWF (2009) Mega-stress for mega cities. A climate vulnerability ranking of major coastal cities in Asia. https://www.alnap.org/help-library/mega-stress-for-mega-cities-a-climate-vulnerability-ranking-of-major-coastal-cities-in. Accessed 1 Mar 2021

Chapter 8
The Path to Net Zero

8.1 The Concept of Net Zero

Climate scientists have calculated the amount of warming that will take place because of cumulative anthropogenic emissions to the atmosphere of carbon dioxide (CO_2) since the beginning of the industrial revolution. This amount is called the "total carbon budget." The more the budget is exceeded, the higher global warming will be. To meet the objectives of the Paris Agreement, the nations of the world need to decrease annual CO_2 emissions to "net zero" by 2050 because the budget of CO_2 emissions that will keep warming to 1.5 °C will be fully exhausted by that date.

In 2020 total annual emissions of carbon dioxide (CO_2) amounted to approximately 42 Gigatons (Gt). Annual emissions of all greenhouse gases (GHGs), including CO_2, was estimated at approximately 59 Gt. With population growth and increases in gross domestic product (GDP) this number had risen steadily since international actions to combat global warming got underway in the late 1990s.

According to the IPCC, "net zero" is achieved "when anthropogenic CO_2 emissions are balanced globally by anthropogenic CO_2 removals over a specified period." In other words, economies have decarbonized energy systems, transport, agriculture, and other sectors to such an extent that any remaining residual CO_2 emissions in a year can be offset by biological removals (growth in forest carbon stocks, for example) or by carbon capture and sequestration. Net zero, then, represents a near complete decarbonization of the worlds' economies in the thirty-year period between 2020 and 2050 (IPCC 2018, p. 24).

Whether "net zero" and "carbon neutral" are synonymous terms is a topic that divides experts. According to ISO 14021:2016 as amended, "'carbon neutral' refers to a product (as a product system) that has a 'carbon footprint' of zero or a product with a 'carbon footprint' that has been offset (ISO 2021, 7.13.3.1)." This means that any carbon polluting activity (e.g., steel production, airplane operation, livestock raising) could be considered "carbon neutral" if the organization responsible for the emissions purchased offset credits in the same amount as its GHG emissions.

J. C. Shideler and J. Hetzel, *Introduction to Climate Change Management*,
Springer Climate, https://doi.org/10.1007/978-3-030-87918-1_8

The French Agency for Environment and Energy Management (ADEME) believes otherwise. The agency expressed its views in an article published in April 2021 with the title "All actors must act collectively for carbon neutrality, but no actor should claim to be carbon neutral." ADEME states that carbon neutrality can only be claimed at a planetary scale or the scale of a nation, rather than at the organizational or product level. Instead, individual actors need to commit to ambitious climate goals designed to fulfill the Paris Agreement in conformity with national objectives. This requires drastic reductions in GHG emissions on the one hand and on the other hand important increases is sequestration of carbon. Quantifying a facility's direct emissions and energy indirect emissions does not account for the totality of indirect emissions associated with the organization or product and therefore produces misleading results. Moreover, purchasing carbon offsets—especially those of low quality—is a strategy that avoids the harder work of reducing one's own GHG emissions and mitigating the impact they have on global warming (ADEME 2021).

Chapter 4 described mitigation activities undertaken by organizations and project developers in the first two decades of the twenty-first century. It reviewed project types approved by the Clean Development Mechanism Executive Board and other approvers of methodologies and protocols. Through voluntary and regulatory programs, organizations and project developers have reduced GHG emissions and enhanced CO_2 removals by several billions of tons since the early 2000s.[1]

Mitigation actions taken through the first two decades of the twenty-first century are insufficient, and offsetting is not the same as decarbonizing. The climate challenge of the third, fourth and fifth decades of the twenty-first century is to limit warming to 1.5 °C by 2050. Achieving "net-zero" emissions means that economic activities no longer emit CO_2, or if they do, in such small quantities that these residual emissions can be offset by such technologies as direct air capture that removes carbon dioxide from the atmosphere or by agricultural and forestry practices that increase net sinks of CO_2 in soil and trees.

The Paris Agreement of 2015 provided an institutional framework for countries to determine national commitments to greenhouse gas emission reduction (see Chap. 2). In this chapter, we describe some elements of a transition strategy to a low-carbon future that are necessary to achieve the Paris Agreement's goal of limiting global temperature rise to no more than 1.5 °C.

---

[1] Authoritative figures can be difficult to obtain. The CDM has issued approximately 3.0 billion tons of CERs since the inception of the program (UNFCCC 2021). Five voluntary registries (Verra, Gold Standard, Climate Action Reserve, American Carbon Registry, Plan Vivo) together have issued approximately 1 billion of tons (Turner and Grocott 2020). The California Air Resources Board has issued approximately 195 Mt offset credits since program inception as of March 2021, Quebec 4.1 Mt CO_2e since 1990, and Ontario 14.4 Mt CO_2e.

8.2 Energy

The dominant source of energy during most of the twentieth century was fossil fuel, primarily coal, oil, and natural gas, the extraction and combustion of which is largely responsible for the global warming from which planet Earth currently suffers. It can be noted that early in the twentieth century, when cities were newly introduced to electricity, urban and suburban transport was frequently electrified. Streetcars and electric vehicles enjoyed spurts of popularity. Cheap oil prevailed, however, creating the problem that societies in the twenty-first century must now address on an urgent basis.

Electrification and the weaning of societies from their dependence on fossil fuels are key to the deep decarbonization that the world needs to achieve by 2050. Abundant supplies of electricity are essential for the "digital economy" as well as for lighting homes, powering factories, and moving people in the "real economy." According to one estimate, decarbonizing the power grid in the USA alone would require tripling the amount of new electricity generation capacity to around 75 GW every year until 2050 (Gates 2021, p. 80).

Advances in the deployment of renewable sources of electricity have been made in the first decades of the twenty-first century, and this trend needs to continue. But even scaling the use of solar, wind, and hydro will not be enough. Marshaling more renewable sources of electricity will also require investment in new high-voltage transmission lines to move electrons from areas of reliable sources of sun and wind to urban centers of demand. As our economy relies more on electricity, system reliability will become ever more important (Gates 2021, pp. 81–82). Increasing grid capacity and transmission will also require consideration of social equity and minimization of adverse negative impacts associated with each deployed technology.

A policy tool that has proven effective in limiting emissions in the electricity power generation sector is "demand-side management." Electric utilities may issue partial rebates to consumers who purchase appliances certified by EnergyStar and similar programs. By reducing energy consumption of refrigerators, for example, residential demand for electricity is reduced, and utilities can serve other electricity needs with the same amount of generation capacity. "Smart metering"—the ability of electricity providers to obtain more detailed information about electricity consumption at each household—used in conjunction with carbon energy production can result in better load balancing at the utility level of local solar, wind, hydrogen, or biomass energy resources.

Information Technology (IT) can contribute to pathways to net-zero energy by informing decisions about energy sourcing and dispatching. Smart technologies monitor both the sources of energy production and its consumption, including distributed energy sources such as residential or commercial solar rooftop panels. Aligning decisions about production and distribution of energy based on the sources' CO_2 emissions rather its kilowatt hours is a task IT is well suited to handle. Incentives to producers should have the effect of increasing low-carbon energy production such as solar, wind, hydraulic, and green hydrogen, and compensation for delivery of electricity should encompass the life cycle costs of energy production including the costs of waste and the end-of-life of equipment and plants. In such reformed markets, an explicit cost of carbon can provide the incentives needed to develop and deploy energy storage solutions that address the intermittency challenge of some renewable energy sources.

8.2.1 Renewable Sources of Electricity

Carbon-free energy from onshore and offshore wind, and solar generation, both at utility scale and distributed energy from rooftop installations, is key to decarbonizing electricity supplies (Gerrard and Dernbach 2018, p. 47). A great advantage of renewable energy is that the sun and wind energy sources are free. Solar energy arrays can be developed at utility scale. Distributed solar energy on rooftops of homes and businesses decentralizes this source of power and enables many consumers to be "energy independent" with respect to their electricity needs. Surplus energy generated can be sold to the grid operator for distribution to other users. The main drawback to solar and wind generation is its intermittency. The sun does not shine 24 hours per day, and winds are sometimes calm. Electricity storage using lithium-ion batteries is currently expensive, and seasonal variations pose an even greater problem than days turning to nights (Gates 2021, pp. 76–79).

One solution to the intermittency problem for renewable energy sources is called pumped storage hydropower. It works by pumping water to a reservoir when renewable sources produce excess energy. The water is then released when needed to power downslope turbines. The Snowy Hydro 2.0 project in Australia, at a capacity of 2 GW, will be the largest power station of this kind in the southern hemisphere when it becomes operational in 2025. According to the Australian National University, more than 600,000 potential off-river sites have been identified globally for this technology, and nearly 300 pumped storage hydropower projects with a planned capacity of more than 200 GW were under construction at the end of 2019 (Turnbull 2021).

8.2.2 Hydroelectric Sources of Electricity

Hydroelectric power has been tapped by societies for centuries. Flour mills located alongside rivers powered the millstones that ground local grain. With the dawn of the era of electricity in the late nineteenth century, rivers were tapped first to provide low-head electricity generated by turbines at the bottom of falls and later by massive hydroelectric dams. In the USA during the Great Depression of the 1930s, large infrastructure projects delivered power to the southwest from Boulder Dam on the Colorado River and to the northwest from the Grand Coulee dam on the Columbia. Lesser projects proliferated.

Hydroelectric development has proceeded around the world. More recent large projects include the Three Gorges dam on the Yangtze River in China (2008) and the Grand Ethiopian Renaissance Dam still under construction in 2021. But dams are not without critics. Large hydroelectric developments can submerge upstream communities and farmlands and displace thousands of impacted residents. Moreover, still lakes created behind dams become anaerobic and generate methane (CH_4) emissions. Dams also impede fish from migrating upstream and impact local populations who rely on fishing. Entire cities such as Seattle and Vancouver, British Columbia, derive most of their electricity from hydroelectric generation at dams. However, this resource is limited and is not likely to expand significantly beyond the installed capacity that it had at the beginning of the twenty-first century's third decade.

8.2.3 Green Hydrogen

Infinitely available and potentially green, hydrogen promises to deliver an ever-growing share of the world's energy mix. Green hydrogen is made through a process known as electrolysis. An electrolyzer splits water into hydrogen and oxygen using an electric current. If the electricity used comes from renewable sources, like wind and solar, the subsequent hydrogen is called "green." According to the International Energy Agency, less than 0.1% of hydrogen today is produced through water electrolysis, but that could soon change.

The barrier has always been the cost of delivering green hydrogen. "We actually see a complete dive down of hydrogen production cost," said Haim Israel, global strategist and head of thematic investing at BofA Securities. He said that electrolyzer prices are down 50% in the last five years, and renewable energy costs have fallen 50–60%. "We believe both of them will go down another 60–70% before the end of the decade" (Petrova 2020).

Experts think hydrogen can be especially effective when it comes to long-haul trucking and even long-haul air travel where using heavy batteries is not feasible (Petrova 2020). Mitsubishi Power and fuel storage company Magnum Development are working on a project in Utah to build a storage facility for 1000 megawatts of clean power, partly by keeping hydrogen in salt caverns. Scheduled to be operational by 2025, the Advanced Clean Energy Storage Project would be the largest clean energy storage system in the world. In fact, BofA Securities analysts think that by 2050 clean hydrogen could account for an estimated 22% of our energy needs, up from just 4% of the energy that hydrogen supplies today. But this would require massive amounts of additional renewable electricity generation (Petrova 2020).

BloombergNEF estimates that generating enough green hydrogen to meet a quarter of our energy needs would take more electricity than the world generates today from all sources combined, and an investment of $11 trillion in production, storage, and transportation infrastructure (Petrova 2020). In the best-case scenario, green hydrogen can already be produced at costs competitive with blue hydrogen today, using low-cost renewable electricity, i.e., around $20 per megawatt-hour (MWh).

Blue hydrogen uses the traditional techniques for producing hydrogen but adds carbon capture and utilization for the CO_2 (IRENA 2020).

8.2.4 Geothermal Sources of Electricity

Geothermal is probably the least well known of all renewable energy. Roman baths, after which the town in Bath in the UK is named, are renowned. Founded in the seventh century, the town of Bath was reorganized in the tenth century and rebuilt in the twelfth and sixteenth centuries. The drains set up in Roman times still carry steaming water from Bath's hot springs to the River Avon. Around 1.1 million liters of hot water flow through the Roman baths from the on-site spring. The used water then flows into the Avon (Richter 2019).

The thermal energy contained in the rocks and fluids under the Earth's crust can be found anywhere from shallow ground to several miles below the surface. It can even be found much farther down in the magma level of the Earth. Geothermal energy can be used to cool buildings or homes and even generate electricity. Some geothermal power plants use the steam from a reservoir to power a turbine or generator, while others use hot water to boil a working fluid that vaporizes and spins a turbine. Geothermal energy is currently only developed in some western USA states such as California, Alaska, and Hawaii. Geothermal energy can also be used for growing plants in greenhouses, drying crops, heating water in fish farms, and in many industrial processes.

Scientific studies classify geothermal energy as a renewable energy, but this energy does not have zero GHG emissions. While a large majority of installations draw their geothermal energy from geothermal reservoirs with low GHG concentrations, the search for low emission energy sources has steered attention to a wider range of geothermal resources. These include geothermal systems with relatively high GHG concentrations in the reservoir fluid. There is a growing realization within the geothermal community that geothermal power plants can, in some cases, release significant quantities of GHGs into the atmosphere.

GHGs are naturally present in all geothermal fluids, and thus geothermal power production from intermediate to high-temperature geothermal resources leads to some GHG releases into the atmosphere. The dominant noncondensable gas (NCG) in geothermal fluids is CO_2, typically constituting more than 95% of the total NCG content. The other relevant GHG in geothermal fluids is CH_4, whose concentration is generally a few hundredths to a few tenths of a percent by mass but in rare cases can make up more than 1.5% of the total gas (i.e., amounting to more than 30% of the GHG emissions as CH_4 traps thermal radiation more efficiently than CO_2). However, most available data on GHG emissions from geothermal power plants refer to CO_2 only (Bonafin 2019).

In 2001, the global average estimate for operational GHG emissions from geothermal power production was 122 gCO_2/kWh, based on a survey involving emissions from power plants that constitute more than 50% of the geothermal capacity installed worldwide. Available data from the USA and New Zealand are consistent with these global emission values, resulting in average figures of 106 gCO_2/kWh (in 2002) and 123 gCO_2e/kWh (in 2012), respectively. The countrywide weighted average emission estimate for Iceland is lower at 34 gCO_2/kWh (in 2013), and the corresponding value for Italy is higher at 330 gCO_2/kWh (in 2013) (World Bank 2016, p. 2).

Natural geothermal produces CO_2 which needs to be captured. Some researchers have developed projects to address this issue. The idea: CO_2 that is cycled through hot regions kilometers underground can efficiently bring heat to the surface, where it can be used to generate electricity. The likelihood is that the process would leave lots of CO_2 underground, and thus out of the atmosphere, according to Symyx project leader and materials scientist Miroslav Petro. "You're sequestering CO_2 and at the same time generating power from it". Backers of this as-yet-unproven concept secured a big endorsement and much-needed cash with a U.S. Department of Energy award of \$338 million in federal stimulus funds for geothermal energy research. Some \$16 million of the funds will be shared by nine CO_2-related projects led by Lawrence Berkeley National Laboratory and other national labs, Sunnyvale, CA-based combinatorial chemistry firm Symyx Technologies, and several US universities (Fairley 2009).

8.2.5 Energy Efficiency

To complement decarbonized energy sources, economies will need to make more efficient use of available energy sources. Based on modeling developed by the International Energy Agency (IEA), if the world were to implement all the cost-effective energy efficiency measures, based on existing technology, energy-related GHG emissions would peak between 2020 and 2040 (IEA 2018). According to this projection, detailed in the IEA's Efficient World Scenario, energy efficiency could deliver a reduction in annual energy-related emissions of 3.5 Gt CO_2-equivalent (CO_2e) (12%) compared with 2017 levels, delivering more than 40% of the abatement required to be in line with the Paris Agreement. Combined with renewable energy and other measures, energy efficiency is therefore indispensable to achieving global climate targets (IEA 2018).

8.2.6 Transitioning Away from Fossil Fuels

Renewable sources of energy have demonstrated their functional equivalence with their fossil-based counterparts. The challenge for the energy sector is to scale up renewable "drop-in" fuels for use in transport vehicles during the period in which fleets are converted from gasoline and diesel fuels made from petroleum while at the same time scaling down exploration and production of legacy fossil fuels. The task is as urgent as it is daunting. Oil and gas companies will need to ramp up participation in the renewables sector, abandon plans to search for new oil and gas fields, and responsibly shut down production of existing oil and gas properties. While this happens consumers will need continued access to legacy fuels at the same time as electric charging stations become more available, electricity production increases to meet higher demand, and fleets are electrified.

May 2021 may be recorded as a turning point for the oil and gas sector. During that month, the International Energy Agency issued a report stating that oil and gas companies need to cease new exploration projects starting in 2021 if the world is to achieve the ambition expressed in the Paris Agreement to limit global warming to between 1.5 and 2 °C (IEA 2021, p. 101). In the same month, the shareholders of ExxonMobil elected three new independent directors who advocate for starting the company's transition away from fossil fuels, the French oil major Total rebranded itself as "Total Energies", and Shell Oil suffered an adverse ruling in a Dutch court that requires it to cut its GHG emissions 45% on an absolute basis by 2030. The judge in the case ruled that Shell had violated the human rights of persons affected by climate change (Raval 2021).

8.2.7 Nuclear Power

Nuclear fission as a carbon-free technology for generating electricity is an existing technology that currently contributes approximately 20% of baseload electricity generation in the USA and about 70% in France. It is highly controversial for many reasons. The first reason is its association with weapons of mass destruction and the threat of their proliferation to irresponsible state and non-state actors. Governments are required to exercise strict controls on the fuel cycle and the operation and security of nuclear-generating plants.

The world remains divided between states with declared nuclear weapons (the USA, Russia, France, United Kingdom, Israel, China, India, Pakistan, North Korea) and those without. Significant military and diplomatic efforts are expended to limit the spread of nuclear weapons to "rogue" states and terrorist groups, but threats remain. And, although the declared nuclear weapons states encouraged non-nuclear weapons states to join them in a nuclear non-proliferation treaty, there is little evidence that the declared nuclear weapons powers will honor the treaty's goal of eliminating nuclear weapons anytime in the foreseeable future.

Disposing of spent nuclear fuel is also a problem associated with nuclear power. In the typical nuclear power plant fuel cycle, mined uranium is made into "low-enriched" pellets that are placed in rods used in the controlled nuclear fission reaction that generates the heat that creates the steam that drives electricity generation turbines. After several years in service, the pellets become "spent" and are replaced by fresh fuel. Spent pellets remain hot and radioactive. They are stored in sealed containers in cooling pools on the grounds of nuclear reactor sites. Since the 1970s, countries have debated the best way of dealing with this nuclear "waste."

Industry's preferred option is to transfer containers of spent fuel to geologic waste repositories where it can remain buried for thousands of years. This solution poses several issues. Will the geologic sites safely contain this radioactive waste? Can the waste be safely transported from power plants to the repository? Will communities surrounding the proposed nuclear waste sites accept their presence? Finally, another solution to the "waste problem" involves reprocessing the spent nuclear fuel to extract plutonium which can then be made into new fuel pellets. This technology, which has been demonstrated in several countries, poses nuclear weapons proliferation risks because plutonium is a prime ingredient in the manufacture of nuclear bombs.

The third problem for deployment of nuclear power plants is economic. Nuclear power plants cost billions of dollars to build, and cost-overruns and construction delays are common. Once built, operations and maintenance costs are reasonably low, but capital costs for siting, permitting, constructing, commissioning and provisions for waste management and decommissioning make new plants economically non-competitive with other electricity generation sources. Governments traditionally have provided nuclear power plant operators generous subsidies to offset these costs.

The final problem facing nuclear power supporters is safety and the public's perceptions of nuclear safety. Fortunately, accidents at nuclear power plants are relatively infrequent. But they can be catastrophic when they occur. Examples include Chernobyl in Ukraine in 1986 and Fukushima in Japan in 2011. In both cases, the cores of generating stations melted down and released large amounts of radioactive material. Inhabited areas surrounding the reactor sites had to be evacuated and will remain contaminated for hundreds of years. Advocates for nuclear energy point to health impacts in the coal mining industry and argue that more miners have been killed or incapacitated in coal mine explosions and from black lung disease than the number of workers who have been harmed from mining uranium or operating nuclear power plants. These arguments rarely persuade persons who are concerned about safety in the nuclear power sector.

The future role for nuclear energy in a decarbonized economy remains hotly contested. The marginal costs for continuing to operate existing nuclear power plants that provide baseload electricity generation are relatively low. Whether the nuclear industry should expand to meet growing demand for electricity will be decided by societal preferences. Industry representatives tout the advantages of latest-generation "inherently safe" reactors (Maykuth 2019). In the meantime, the IPCC projects that nuclear energy as the source of primary energy supply will grow from 10.91 exajoules (EJ) in 2020 to 16.26 EJ in 2030 and 24.51 EJ in 2050 (IPCC 2018). It is likely that this more than doubling of installed nuclear reactor generating capacity will require a much more profound shift in both economics and public acceptance than has been demonstrated during the industry's history since 1960.

8.2.8 Natural Gas-Fired Generating Stations

Natural gas-fired generating stations, when equipped with carbon capture technology, may play a limited role in a net-zero future. Their chief advantage is that they are fast-starting and capable of balancing peak electrical loads quickly when other generation sources are inadequate to meet demand. Renewable energy sources such as solar, wind, biomass, hydroelectricity, or geothermal now compete favorably with natural gas on economics at utility scale and need only to solve the problem of intermittent generation to serve as a full substitute for fossil fuels.

8.2.9 Renewable Energy as a Fossil Fuel Replacement

Renewable energy has officially moved from the concept stage to implementation. According to a Global Trends in Renewable Energy Investment 2018 report, global investment in renewable energy topped $240 billion for the eighth straight year in 2017. In 2017, $279.8 billion was invested (Frankfurt School of Finance and Management 2018).

In 2018, more than $1 billion was allocated to research and development projects targeting renewable energy. Notably, 29 countries committed funds to renewable energy projects in 2018 compared to just 21 in 2017. These investments have led to an unprecedented expansion of the global renewable energy industry which in 2017 was valued at approximately $928 billion. Further technological improvement, as well as increasing avoidance of fossil fuels, should encourage more growth in this sector.

China is by far the largest player in the renewable energy industry. In particular, China had an installed renewable energy capacity of 695.87 GW in 2018. This is as much capacity as the next four countries combined. China consumed 143.5 million tons of oil equivalent (Mtoe) in 2018 compared to 103.8 Mtoe for the USA, 47.3 Mtoe for Germany, and 27.5 Mtoe for India (Gupta 2019).

The heating and cooling sector accounts for 51% of renewable energy consumed. Consumers, for example, use solar energy to heat water and for space heating. The transport sector accounts for another 31% of renewable energy use. The proliferation of electric vehicles is the driver of this high consumption rate (REN 21 2020).

Manufacturing industries and other corporations are current consumers of renewable energy. Companies like Apple, Walmart, and Target have installed rooftop solar panels to generate electricity for their operations. According to the 2018 Solar Means Business Report, more than 35,000 projects in 43 US states have an aggregated installed solar capacity of 7000 MW. Apple had the highest installed solar capacity at 393.3 MW (Gupta 2019).

The declining cost of renewable energy generation is the biggest factor driving growth in the industry. As more companies continue to invest in technological advancements, costs should fall further. For instance, in nine years after 2010, it costs 79% less to store electricity in a battery. And the price of solar and wind power production was 18% cheaper in 2019 compared to 2018.

Looking forward, supportive policies from the government should spur further growth in the industry. Investors have discovered the sector, a signal that more funds will be invested to make renewable energy more affordable in the future. To cap it all, advancing technologies have made it easier and cheaper to generate renewable power, hence pushing down its cost to consumers (Gupta 2019).

The EU taxonomy supports the sector with metrics and thresholds that state that "any electricity generation technology can be included in the taxonomy if it can be demonstrated, using an ISO 14067 or a GHG Protocol Product Lifecycle Standard-compliant Product Carbon Footprint (PCF) assessment, that the life cycle impacts for producing 1 kWh of electricity are below the declining threshold." In the EU taxonomy the declining threshold refers to facilities whose operating emissions are lower than 100 gCO_2e/kWh, declining to net-0 gCO_2e/kWh by 2050 (European Commission 2020a).

8.3 Manufacturing

Manufacturing is the second largest contributor to CO_2e emissions. The sector also makes the products and technologies that can contribute to GHG emission reductions in other sectors of the economy and is thus a fundamental part of the low-carbon economy. The manufacturing sector provides support to those economic activities that are low in carbon emissions and first movers who are engaging in a trans-formational shift (European Commission 2020b). "Enabling" activities cover both those included under "low-carbon technologies" and "mitigation measures" which when combined result in GHG emission reductions. Low-carbon activities refer to the manufacturing of products, key components, equipment, and machinery that are essential to a number of key renewable energy technologies (geothermal power, hydropower, concentrated solar power (CSP), solar photovoltaic (PV) technology, wind energy, and ocean energy); the manufacturing of low-carbon transport vehicles, fleets and vessels; the manufacturing of energy efficiency equipment for buildings and other low-carbon technologies that result in substantial GHG emission reductions in further sectors of the economy (including private households) (European Commission 2020a).

8.4 Transport

To achieve climate neutrality, a 90% reduction in transport emissions is needed by 2050 (compared to 1990). Road, rail, aviation, and waterborne transport will all have to contribute to the reduction. Currently, transport operations consume one-third of all energy. The bulk of this energy comes from oil. This means that transport is

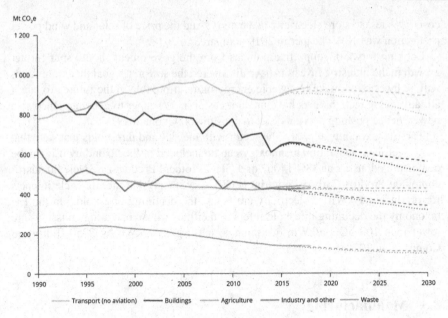

Mt CO₂e

Fig. 8.1 Trends and projections in Europe 2019 (EEA 2019)

responsible for a large share of GHG emissions and is a major contributor to climate change. While most other economic sectors, such as industry, have reduced their emissions since 1990, those from transport have risen (see Fig. 8.1). In Europe, GHG emissions from transport now account for more than one quarter of total GHG emissions. Transportation thus presents a major challenge for ensuring that the emission reduction targets are met. Although vehicle efficiency improvements have had a mitigating effect on GHG emissions, growing transport demand and a sluggish share of low-carbon solutions have hampered decarbonization progress.

Within the transport sector, road transport is the dominant emissions source in Europe, accounting for more than two-thirds (71.7%) of transport-related greenhouse gas emissions. Passenger cars and vans are responsible for the bulk of these emissions, with the rest resulting from trucks and buses. Road transport is followed by shipping and aviation as the second and third largest sources of GHG emissions from transport. The transport sector represents about 30% of additional annual investment needs for sustainable development in the European Union (European Commission 2020b).

In the USA, the transport sector accounts for 28% of GHG emissions (Fig. 8.2).

The principal options for climate mitigation in the transport sector include increasing the number of low- and zero-emission vehicles and improving vehicle efficiency and infrastructure. A long-term strategy for fleet efficiency and fuel substitution involves three categories of transformation.

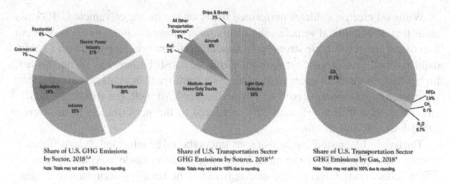

Fig. 8.2 Share of US transportation sector GHG emissions 2018 (US EPA 2020)

8.4.1 Efficient, Low-, and Zero-Direct Emissions Fleets

Vehicle fleets become more efficient over time by meeting increasingly stringent emissions performance standards set in regulations. These policies ensure substantially reduced emissions. Performance metrics (vehicle emissions per km traveled [vkm], number of passengers moved per km [pkm], or freight tons transported per km [tkm]) are mode-specific and are linked to available testing methods where available. They require efficiency improvements without being technology prescriptive, so long as the benefits of relevant technologies can be demonstrated.

A life cycle assessment (LCA) study described by Carbon Brief (2019) compared GHG emissions from conventional cars to those of electric cars operated in different countries (Fig. 8.3).

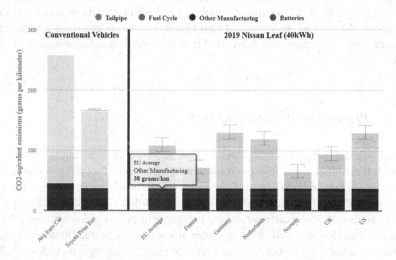

Fig. 8.3 Comparison of conventional and hybrid and electric car performance including the variation of electricity-related GHG emissions in different countries (Carbon Brief 2019)

While all electric vehicles performed better on the metric of vehicle GHG emissions than conventional vehicles, the comparison showed that the location of operation of the electric vehicle strongly influenced performance. Norway, with its abundant hydroelectric electricity, reduced GHG the most. France was second with its high reliance on nuclear energy. Countries with significant amounts of thermal energy generation like Germany and the USA performed least well, but still reduced GHG emissions by a little more than half compared to the performance of an average fossil-fueled European car.

The results of the study show that consumers should be informed about differences among vehicle types. For example, a Tesla using a battery made in Asia only achieves a 50% reduction in GHG emissions compared to the better performance of a lighter car using a battery made in the UK. This highlights the need to make product-specific information available at points of sale to consumers who may be seeking to reduce their transportation climate change footprint.

8.5   Mining/Mineral Production

This is an important sector both in terms of avoiding bottlenecks in the deployment of low-carbon technologies that provide the critical materials needed for low-carbon technologies, and for the value chain link with energy-intensive manufacturing sectors. LCA should be applied to understand the full impacts of metals used in the manufacture of low-carbon products. Aluminum helps produce lighter weight cars. Copper is used for electrical wiring and in electric vehicle motors. Aluminum and copper are also used extensively in the manufacturing of solar panels and wind turbines. Metals used in batteries, such as cobalt, lead, lithium, manganese, and nickel, are key for low-emission mobility and for electricity grid storage. Zinc and cobalt are used in offshore wind turbines. Silicon is an important raw material for solar panels. Precious metals are used in both vehicle manufacturing and in solar panels.

8.5.1   Production of Steel and Iron

In the long term, the iron and steel making industry aims to implement breakthrough technologies to achieve ultra-low CO_2 emissions. Promising technologies have already been demonstrated at the pilot or at industrial scale. A Swedish firm has produced direct reduced iron (DRI) in a process that uses hydrogen to separate iron from its ore. To be carbon-free, the process relies upon green hydrogen that is produced from Sweden's abundant hydroelectric resources. In traditional processing, iron is extracted from its ore in a blast furnace using metallurgical grade coal. The traditional process is responsible for as much as 9% of the world's greenhouse gas emissions. After using DRI to extract iron from ore, steel can be made in electric arc furnaces that do not produce any process emissions of CO_2 (Milne 2020).

Electric arc furnaces contribute to the production of low-carbon steel by using recycled steel as its feedstock. Further climate change mitigation gains can be made by increasing the recyclability of manufactured steel products. The use of high-strength lighter weight steels can also improve the carbon footprint of this important sector.

8.5.2 Production of Aluminum

Electricity is one of the major inputs to the manufacture of aluminum. Producers with access to renewable electricity are at a competitive advantage in the quest for reduced emissions from this sector. The EU Technical Expert Group (European Commission 2020a) that advised the European Union on the development of its taxonomy for sustainable investments made the following observations about aluminum:

– Electricity costs typically contribute to more than 50% of the production costs; consequently, there is a strong incentive for the aluminum industry to improve its energy efficiency.
– The key action for aluminum production to make a substantial contribution to climate change mitigation is to increase its use of low-carbon electricity.
– It is acknowledged that aluminum production facilities can play an important role in stabilizing electricity grids by active management of electricity demand; this can result in substantial mitigation contributions, e.g., by limiting the need for electricity storage facilities.
– Aluminum will play an important role in the low-carbon economy, enabling lightweight products and electrification (including transmission wires with low conductivity losses).
– Aluminum recycling and the manufacture of aluminum products with high recycling can make a substantial contribution to climate change mitigation because of its association with much lower emissions than primary production.

8.6 Chemical Industry

The "green chemical industry" covers the production of bioalcohols, biopolymers, biosolvents, and organic acids used in sectors as diverse as food and beverage, personal care, packaging, automobile, and agriculture. Valued at $9.5 billion in 2019, the global green chemicals market size is projected to grow at a compound annual rate of 6.6% during in the period 2020–2030. Rising acceptance of biobased packaging, the harmful effects of inorganic chemicals, and increasing concerns regarding fossil fuel depletion are factors driving the growth of the green chemicals industry (Prescient and Strategic intelligence 2020).

In 2019, the bioalcohol category accounted for the largest market size by product type in the green chemicals sector. This is explained by the adoption of bioethanol in the production of transportation biofuels, along with the increasing usage of bioalcohol as an intoxicating agent in alcoholic beverages. Bioalcohols are favored by governments in several countries to reduce the formation of ground-level ozone, a constituent of smog (Prescient and Strategic intelligence 2020).

In 2017, the Joint Research Centre (JRC)—an EU Commission research group—found that the chemical and petrochemical sector can support a 45.6% increase of production and, by implementing the best available and emerging energy technology, could achieve 4% energy savings and reduce GHG emissions by 75.5 Mt CO_2e (a 36% reduction) by 2050 (European Commission 2017).

With the output of each battery production facility measured in terawatt-hours, chemical energy storage maintains a leading position in the energy storage sector. It is also the only option for seasonal energy storage using the charging technology power-to-gas in combination with the existing gas infrastructure for storing and converting gas into electricity.

Energy stored in the form of hydrogen (H_2) or methane can be used by all three sectors—electricity, heating, and transport. There is already a large existing infrastructure for transporting, distributing, and using methane in gas-fired boilers and in combined heat and power (CHP) facilities; in gas vehicles and ships using LNG in the transport sector; and in gas turbines, combined cycle gas turbine plants, and CHP in the electricity sector. The infrastructure for hydrogen is still under development, and currently only available in industrial parks. Surplus energy from renewable energy sources can be temporarily stored in the gas network or in gas storage facilities and then supplied to other locations when demand is higher.

Only chemical energy storage can combine energy storage and energy transport at this scale. The transmission capacity of a large gas pipeline is about ten times greater than that of a high-voltage transmission line. There is also significantly greater public support for expanding the gas network than for expanding the electricity network. But electricity transmission is more efficient than converting electricity into gas and then transporting the gas (Stadler and Sterner 2018).

8.7 Construction

Energy use in buildings is a significant contributor to the world's GHG emissions. We noted in Chap. 4 the limited abatement results to date from the implementation of CDM emission reduction projects. In this chapter, we look to the future but begin with the observation that new buildings constitute less than 1% of the total number of buildings constructed. If decarbonization focused only on new buildings, it would take one hundred years to turn over existing building stocks if all new buildings conformed to green building standards. To meet the decarbonization challenge of the first half of the twenty-first century, existing buildings must be refurbished to correct their lack of insulation, excessive energy demand, urban soil impermeabilization, and low quality of construction.

The energy used in the construction and operational stages of buildings is responsible for more than one-third of total global GHG emissions—11% in building material and construction and 28% in building operation (Ali et al 2020). The performance of buildings should be analyzed using life cycle assessment over expected lifetimes of about 50 years, considering not only the construction stage but also the use and the end-of-life stages. From an economic perspective, higher initial construction costs of green buildings are more than compensated for during a building's expected life by accounting for reduced expenditures on energy during the use stage.

A first principle should be the use of bioclimatic construction to capitalize on a site's natural advantages. Historical architecture points to the best rules for orienting a building. In temperate zones, a building should capture the sun's warming rays in the winter while screening them in the summer. In tropical areas, wind helps circulate fresh air and moderate the sun's warming effects. White colors in North Africa reflect the sun, and doors open toward the north to protect against the Sahara wind (Sirocco). Using local breezes to refresh occupied spaces reduces energy demand during the use stage of the building. This principle is simple and less costly but forgotten by many architects and builders.

A second principle is the use of low-carbon construction materials. Wood-framed homes are quite common in the USA, less in Europe. In Asia, some buildings, including skyscrapers, are built using treated bamboo (CRG Architects 2021). Architects in the UK designed a 35-m-tall building in east London supported by cross-laminated timber (CLT). It was the world's largest CLT building upon its completion in 2017 (Foroudi 2021). Customary steel and glass construction still dominate in city centers, highlighting the need to decarbonize these materials. The use of concrete to build prefabricated houses economizes construction time but points to the need to expand the availability of green cement.

Recycling building materials at the end-of-life stage should be pursued but challenges remain. Wood treatment and coatings can limit recyclability, and concrete from demolished buildings is not normally reused in new building construction,[2] but it can be recycled in roadways and similar civil engineering works. Steel can be recycled to use in the construction of new buildings. The manufacture, transport, and installation of building materials such as steel and concrete require a large quantity of energy, despite them representing a minimal part of the ultimate cost of the building. This energy is said to be "embodied" in the materials used for construction. Wood used in construction is sourced not only from lumber but also particleboard which contains formaldehyde. At present, much wood retrieved from demolished buildings is sent to construction and demolition landfills where the percentage of wood was estimated to be 6–7% of total waste emplacements (Cochran and Townsend 2010).

---

[2] Foroudi reported (2021) that technology developments such as diamond wire saws are making possible the cutting and reuse of elements of concrete buildings.

The environmental impact of concrete and steel, considering their embodied energy, is significant. The embodied energy in one ton of concrete, when multiplied by the huge amount of concrete used, results in concrete being the material that contains the greatest amount of carbon in the world. The embodied energy of concrete is 12.5 MJ/kg, that of steel 10.5 MJ/kg and that of wood 2.0 MJ/kg (Hsu 2010). Architects are beginning to take note.

Wood buildings, on the other hand, require much less energy from resource extraction through manufacturing, distribution, use and end-of-life disposal, and are responsible for far less greenhouse gas emissions, air pollution, and water pollution (Evan 2016). The construction sector has commissioned research to find ways to reduce GHG emissions or capture CO_2 to mitigate the effects of embodied energy. Valid studies take the entire life cycle of a building, including its use and end-of-life stages into account when assessing environmental impacts. The government of France has decreed that from 2022 onward all public buildings should be made with at least 50% wood (Foroudi 2021).

A third principle is to insulate a building's external skin. This approach is common in the USA and has gained adherents in Europe for new green buildings. Enhanced insulation should be used for all new buildings, and weatherization techniques can be implemented on existing building stocks. With local production of renewable energy, it is possible that a building can produce more energy than it consumes for heating, cooling, and lighting (Hetzel 2003). This objective should be extended to all new and refurbished buildings.

A fourth principle is to reduce the extent of impermeabilization of soils in urban area which accelerates floodwaters and raises urban temperatures through the "heat island" effect. A better system is to use green roofs based on native species to reintroduce biodiversity in the urban area. Green roofs are important elements of green buildings and research, including by NASA, has validated their positive environmental impacts (Dunbar 2013).

The coronavirus pandemic of 2019–2021 prompted many office workers to retreat to their homes for safety during lockdowns ordered by national or local authorities. The pandemic also prompted some workers to abandon cities for more rural locations. This retreat from crowded cities was made possible by broadband Internet connections and videoconferencing platforms which substituted for in-person meetings. Only the passage of time will tell how permanent these changes in working habits will become. In the meantime, the experience of commercial real estate in the London market is instructive.

The flight of London office workers to more spacious dwellings outside the city prompted some speculation that high vacancy rates in office buildings would continue and that as many as 10% of employees would not return to their offices (Hammond 2021). Against this background, the acquisitions of land and announced starts of new office buildings in 2021 seemed counterintuitive to many real estate professionals. The logic became clearer when property developers realized that the market was bifurcating between buildings with access to open air and modern amenities and older traditional office spaces. The former were expected to be in high demand while more obsolete structures were predicted to lose tenants and value. Demand for "sustainable" and "net-zero [GHG emissions]" office space has grown. Property market experts even began using the term "stranded assets" to refer to buildings that failed to meet sustainability criteria (Hammond 2021).

8.8 Fugitive Emissions

The World Bank in its "Zero Routine Flaring by 2030" initiative has sought to limit GHG emissions from associated gas produced during the extraction of crude oil from reservoirs. Much of this gas is utilized because governments and oil companies have made investments to capture it. Too much is unnecessarily flared because of technical, regulatory, or economic constraints. As a result, thousands of flares at oil production sites around the globe burn approximately 140 billion cubic meters of natural gas annually, causing more than 300 million tons of CO_2 to be emitted to the atmosphere. Flaring of gas contributes to climate change and impacts the environment through emissions of CO_2, black carbon, and other pollutants.

Flaring also wastes a valuable energy resource that could be used to advance the sustainable development of producing countries. For example, if this amount of gas were used for power generation, it could provide about 750 billion kWh of electricity, or more than the African continent's current annual electricity consumption. While associated gas cannot always be used to produce power, it can often be utilized in a few other productive ways or conserved (re-injected into an underground formation). The oil and gas industry already conducts flaring to reduce safety risks from crude oil transportation and delivery operations. The World Bank's initiative targets routine flaring rather than flaring for safety reasons or non-routine flaring, which nevertheless should be minimized. Routine flaring of gas is flaring during normal oil production operations in the absence of sufficient facilities or amenable geology to re-inject the produced gas, utilize it on-site, or dispatch it to a market. Venting is not an acceptable substitute for flaring (World Bank 2021).

8.9 Waste Handling and Disposal

There is a potential for increased circular economy innovations in wastewater purification and recycling. The international space station provides a model for wastewater reuse. Municipal solid waste is a valuable resource that can be made into jet fuel. A key challenge for commercializing the technology is sorting and standardizing the feedstock for optimal processing at facilities such as Fulcrum Bioenergy's Nevada plant or at Shell/Velocys' facility near Heathrow in the UK.

Chicago's O'Hare international airport transformed the typical runway replacement process through the implementation of sustainability measures that reduced road transport of imported construction material and recycled used concrete from the old runway.

8.9.1 Carbon Capture and Storage

Carbon capture and storage (CCS) initiatives gained prominence in 1996 when the Norwegian state oil company Statoil (now Equinor) first began separating CO_2 from natural gas produced in its North Sea Sleipner gas field. The company reinjected the CO_2 in an existing well. Similar projects followed in Canada and the USA as energy companies found that the reinjected CO_2 could help repressurize depleted reservoirs and thereby enhance oil and gas recovery. Emergence of the technology raised hopes that CCS could be fitted to thermal power generating stations to remove CO_2 emissions from electricity generation before they were released into the atmosphere and permanently store the CO_2 underground.

Despite more than a decade of development at demonstration plants, CCS has failed to gain industry acceptance due its high energy consumption and unattractive economics. The challenges are even more difficult for projects aiming to retrofit existing power plants than for those intended to include CCS in the design of new-build facilities. As a near-term CO_2 abatement measure, CCS may play a role on the path to net zero if it is made part of a regulatory framework limiting GHG emissions in the power sector or if breakthroughs occur in its underlying technology.

8.10 Afforestation and Reforestation

"Nature-based solutions" are favored by many environmentally conscious investors. Forests are valuable for planet Earth not only for the carbon they sequester, but also for other ecosystem services they provide such as nature walks, pollination of crops by bees, and attenuation of flooding in residential areas provided by riparian buffers and wetlands (Johnston 2018). In addition to removing carbon from the atmosphere, these techniques improve water quality, provide habitat for animals, increase biodiversity, prevent erosion and landslides, and cool local temperatures via increased evapotranspiration. Some studies tout the positive impact of cropping in forest lands based on a temperature reduction of $0.1°$ C as calculated by a climate model (Unger 2014). This solution may help justify cropping in forest lands, but it is also clear that application of this technique alone will have only a limited impact on the continuing increase in global mean surface temperatures which were expected to rise from 1.1 °C in 2020 to 1.7 °C in 2050 assuming fulfillment of 2020 emission reduction commitments with no further ambition achieved (Carrington 2020).

8.10.1 Voluntary Initiatives

The effort to decarbonize economies demands universal participation. Government policy makers have an important role to play, but so do private companies. At the beginning of the 2020s, a growing number of prominent multinational companies have declared their commitments to achieve "net-zero" emissions by 2050 (Hicks 2020). Some companies are adopting "science-based targets" in accordance with Science Based Targets Initiative (2021), others are defining targets based on what appears to represent "best available technology" (Science Based Targets Initiative 2021).

The path to net zero will proceed on different timeframes depending on the economic sector. Companies and governments seeking to make good on net-zero promises will be obliged to invest in mitigation projects that will reduce GHG emissions outside their operational boundaries and those of their supply chain partners. According to the Taskforce on Scaling Voluntary Carbon Markets, voluntary emission reduction and removal enhancement projects will need to increase 15-fold in the period 2018–2030 to meet climate commitments, assuming that carbon credits were used to finance the necessary GHG mitigation actions (IIF 2020).

The Case of Shell

The oil and gas company Shell, in its strategic thinking called "Sky" (Evans 2018), claims alignment with the less than 2 °C trend which it considers "technically and economically feasible." This strategic vision represents a shift from Shell's previous statements on this subject. As recently as 2015, Jeremy Bentham, Shell's vice president for environmental policies and future scenarios, said he had analyses supporting achievable scenarios that would restrain warming to 2 °C, but that the actions needed were politically and socially unacceptable and therefore not publishable. The result would be a policy of reducing the carbon footprint of its operations by 50% by 2050 by investing $1 billion to $2 billion annually in renewable energy out of a budget of $25 billion to $30 billion annually. As the remaining investments continue to support recovery of conventional energy assets, the renewable focus remained largely insufficient.

The new SKY scenarios are based on net total zero emissions for 2070 and extremely low emissions from 2020 to 2070 with a clear message: "The Paris Agreement goal of less than 2 °C scenario is possible." This huge challenge for society is technologically and industrially possible "and we, as a multinational, have chosen that to happen. This is a challenging challenge," says Bentham, which contradicts the above established 2.5–3°C scenarios, even if doubts persist about the political and societal acceptance of the Paris Agreement. The framework for the new perspectives considers that coal demand is on a downward trend, with peak oil to be reached in 2025 and peak natural gas in 2030. The result would be an acceleration of solar, which is expected to become the leading energy source in 2060. There would also be strong growth in bioenergy, wind power, and nuclear power. Electricity would increase fivefold worldwide by 2070, focused primarily on electric transport and heating. The price of carbon is expected to be above $40 per ton in 2030 [Note: the price of European Union Allowances exceeded $60 in 2021] and around $80 in 2040. Most European countries could reach zero emissions by 2060].

To reduce emissions, Shell would invest heavily in carbon capture and storage (CCS) while developing bioenergy with carbon capture and storage (BECCS). Shell is targeting 10,000 industrial sites for CCS (currently there are 3000 coal-fired power plants in the world).

8.10.2   The Role for Standards

It is important that organizations of all sizes and types contribute to efforts to reduce the carbon intensity of their processes and products. In many cases, reducing carbon emissions can also mean saving money on fuel and electricity. For incremental reductions achieved up to 2020, energy efficiency has represented a "win-win" for both the planet and the sustainability of individual organizations. Many gains have been achieved by organizations implementing ISO 14001, Environmental management systems—Requirements with guidance for use (ISO 2015a). This standard shares a common structure with ISO 9001, Quality management systems—Requirements, which is universally accepted as standard practice for ensuring that customer requirements are understood, and that goods and services produced or delivered by an organization meet customer expectations (ISO 2015b). Organizations that operate their businesses in conformity with both standards are likely to be well managed and competitive in their respective markets.

ISO 14001 requires organizations to establish an environmental policy and set environmental objectives. Organizations identify risks and opportunities associated with the context in which it operates, including issues identified by interested parties and environmental conditions that it influences or that affect it. The standard is prescriptive in the sense that it defines what management steps must be taken. It leaves to each organization, however, the job of determining the actions it will take to meet the requirements. The standard addresses such topics as leadership, resources, operations, performance evaluation, and improvement.

Standards are particularly important in national and international trade. As part of their supplier qualification process, customers may require organizations in their supply chain to demonstrate conformity with ISO management system standards. According to the World Trade Organization, such requirements do not constitute a "technical barrier to trade" because ISO standards are developed in accordance with an international consensus-based process and are applied in a non-discriminatory fashion.

Industry will continue to search for new ways to decarbonize. Environmental management systems based on ISO 14001 focus attention on the environmental aspects of each organization. This includes climate change, and environmental professionals should continue to search for alternatives to the use of fossil fuels, increased recyclability, and a more circular economy. Life cycle assessment using ISO 14044 Environmental management—Life cycle assessment—Requirements and guidelines (ISO 2006) and ISO 14067 Greenhouse gases—Carbon footprint of products—Requirements and guidelines for quantification (ISO 2018a) help companies identify environmental impacts, including those associated with climate change mitigation.

The economic transition called for by a low-carbon strategy implies decarbonization over a product's life cycle. Mitigation of supply chain emissions is encouraged by mandatory reporting of indirect emissions as prescribed by ISO 14064 Part 1, Greenhouse gases—Specification with guidance at the organizational level for quantification and reporting of greenhouse gas emissions and removals (ISO 2018b).

ISO 14064-1:2018[3]

Clause 5.2.4

GHG emissions shall be aggregated into the following categories at the organizational level:

(a)   direct GHG emissions and removals
(b)   indirect GHG emissions from imported energy
(c)   indirect GHG emissions from transportation
(d)   indirect GHG emissions from products used by the organization
(e)   indirect GHG emissions associated with the use of products from the organization
(f)   indirect GHG emissions from other sources.

In each category, non-biogenic emissions, biogenic anthropogenic emissions and, if quantified and reported, biogenic non-anthropogenic emissions shall be separated (see Annex D).

The ISO 14064-1 approach to GHG quantification provides accounting tools to support transformational decarbonization of businesses by shining light on the embodied carbon in upstream materials used by organizations as well as the emissions of their own processes and downstream emissions in product use and end-of-life stages. An increased focus on upstream emissions allows organizations to evaluate those that are significant for their operations and work with suppliers to reduce them. Lessons learned from the coronavirus pandemic will also encourage organizations to rethink the need for employees to travel as frequently for business purposes and to require employees to report to offices when productivity can be maintained using remote working arrangements. In this way, the low-carbon economy should reinforce frugality in the use of raw materials and resources to the benefit of the environment, workers, and communities.

ISO is one of hundreds of standards development organizations (SDOs) around the world. SDOs are important because they influence voluntary and regulated behavior through contracts, industry-best practices, and codes and regulations. For example, the non-profit US Green Building Council (US GBC) operates the Leadership in Energy and Environmental Design (LEED) program which certifies green buildings worldwide. The certifications, which may be earned at the bronze, gold, and platinum levels, describe the extent to which best practices on energy conservation and other environmental criteria have been met (US GBC 2021). Like LEED, the UK-based Building Research Establishment Environmental Assessment Method (BREEAM) offers sustainability certification for buildings and communities, infrastructure, housing, and fire and security products and services (BRE Group 2021).

Policy makers use standards to meet environmental objectives. Examples include corporate average fuel economy standards in the USA, energy efficiency standards such as US EPA's EnergyStar label for appliances and computers, and building codes to improve the thermal insulation of newly constructed buildings. Many other countries have similar standards. France, for example, has standards for home building that can result in low-carbon buildings or even buildings that produce more energy than they consume. The main issues in realizing emission reduction objectives in this sector is the lack of application of the standards by builders and the fact that the rate of building replacement is only 1% per year (Criqui et al. 2009).

The inability of the world's economies significantly to bend the curve on GHG emissions since the ratification of the Kyoto Protocol in 2005 means that incremental progress toward reducing emissions is insufficient. Governments need not only to implement strict environmental standards but also embark on ambitious plans to decarbonize the transportation, building, manufacturing, and agricultural sectors.

8.11 Making Economies Circular

As early as 1966 Kenneth Boulding raised awareness of an "open economy" with unlimited resource inputs and outputs of products, co-products, and wastes. This open economy contrasted with a "closed economy" in which inputs and outputs were linked and remain integral parts of the economy. Boulding's essay "The Economics of the Coming Spaceship Earth" is often cited as the first expression of the circular economy, although Boulding does not use that phrase.

The circular economy is grounded in the study of feedback-rich (nonlinear) systems, particularly living systems. The contemporary understanding of the circular economy and its practical applications to economic systems has evolved and incorporates different features and contributions from a variety of sources centering around the idea of closed loops. Some of the relevant theoretical influences are cradle to cradle, laws of ecology (see Barry Commoner *The Closing Circle* (1971)), looped and performance economy described by Walter Stahel (1982), regenerative design, industrial ecology (see Allenby (1999)), biomimicry, and the blue economy.

One influential pole of leadership on the circular economy is the Ellen MacArthur Foundation (Ellen MacArthur Foundation 2015) which advocates for the concept that the by-products or "wastes" of one organization can be valued material or process inputs for another. A growing interest in the circular economy led the International Organization for Standardization to create a Technical Committee with the mandate to write international standards that can be used by industry to describe and develop these concepts.

Industry has adopted economically attractive opportunities designed to reduce waste, increase recycling and reuse, and to remanufacture industrials products. (Printer cartridges are a common example.) The EU's directorate general for the environment promotes circular economy initiatives, particularly in the plastics and waste reuse sectors (European Union 2019).

As important as the circular economy is for resource conservation and waste minimization, implementing circular economy precepts is insufficient in itself to achieve the needed decarbonization of economies. However, while planning for decarbonization and implementing strategies to reduce GHG emissions, the benefits of the circular economy should be pursued as suggested by Fig. 8.4. Indeed, working toward a circular economy is an objective of the European Union and is recognized as such in the Technical Expert Group's final report on the EU taxonomy (European Commission 2020a).

8.12 The Role for Finance on Pathways to a Low-Carbon Future

We provided an overview of finance and its role in greening the economy in Chap. 6. In the following paragraphs, we suggest the role finance will play in the reallocation of financial resources from carbon intensive assets and companies to those with the ability to bring transformational or disruptive low-carbon solutions to the market.

In the early years of the 2020s, some investors divested their shares in oil and gas companies because of their high carbon emissions. Financial markets, whose prices reflect the aggregated expectations of individual investors for future profits, may discount the value of fossil-fuel-linked investments because of fears that oil and gas reserves could become "stranded assets."

In May 2021, the International Energy Agency lent credibility to this fear in a report called "Net Zero by 2050: A roadmap for the global energy sector." In its report, the agency, which is considered conservative by many observers (Dupin

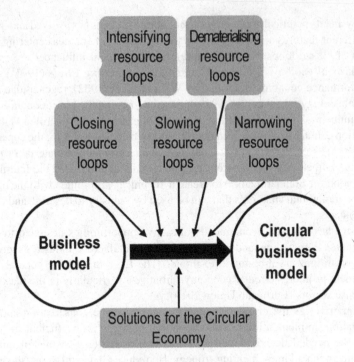

Fig. 8.4 Representation of the circular economy (Jidb175 2018)

2021), said that "no exploration for new resources is required and, other than fields already approved for development, no new oil fields are necessary" if the rise in global surface temperatures is to be limited to 1.5 °C by 2100 (the world reached 1.2 °C in 2020 according to the World Meteorological Organization [Hodgson 2021]).

The report sent an important signal to industry about the energy transition that needs to occur to achieve the decarbonization goals of the Paris Agreement (EIA 2021). That same month the environmental ministers of the Group of Seven most industrialized nations stated in a communiqué that "international investment in unabated coal must stop now" (Hook and Hodgson 2021). Decisions such as these reinforce the notion that an end to widespread crude oil extraction may arrive within 10 years, and natural gas exploitation within 20, no longer seems far-fetched.

Two facts from the EIA report need to be emphasized (Dupin 2021):

1. Economies have all the technologies available to reduce emissions between 2020 and 2030 based on energy efficiency at a rate of 4% per year or three times the reduction achieved in the first two decades of the twenty-first century (1.2%). Methane emissions need to be reduced by 75% over the next few years by exerting tighter control on oil and gas field venting and leak detection. The growth of renewable energies driven by the annual increase in photovoltaic solar will reach 630 Gigawatts (GW) and 390 GW for wind. Production capacity will increase fourfold by 2030, and the sale of electric vehicles will increase 18-fold.

2.   Half of the energy reductions will come from systems still in the form of proto-
     types such as new generations of batteries, hydrogen, and CO_2 capture and
     storage by 2050. Public and private research and development spending will
     need to increase from the current $25 billion to $90 billion by 2030. Electricity
     will account for 50% of total energy consumption by 2050, playing a key role in
     all sectors. Electricity generation will come from 90% renewable energy, wind
     and solar accounting for 70%, nuclear power providing the remainder.

Investments of $5 trillion by 2030 will be needed compared to the current $2
trillion rate, contributing to 0.4% of global GDP growth per year. This IMF-supported
analysis also highlights the capacity to create 14 million jobs by 2030. However, there
is a strong tension in the rare earth and metals markets. Lithium consumption will
increase 40-fold and nickel by 20-fold, and manganese sevenfold by 2040 (Dupin
2021).

The automobile manufacturing sector is facing a similar need for heavy investments to build the plants necessary to bring to market to electric or hydrogen vehicles
and their critical components such as batteries and green hydrogen. The rise of Tesla
and its market capitalization values the electric automobile maker more like a high
technology company than a traditional automaker. Could the food sector also be
disrupted by decarbonization? At least in the developed world, consumer demand
for organic products and plant-based substitutes for dairy products may reflect early
signs of a turning away from products associated with high carbon emissions.

The pursuit of decarbonization will create economic winners and losers. The
thermal coal mining industry will continue to decline. Even metallurgical coal will
cease to find buyers when steel plants have transitioned to direct reduced iron using
green electricity. The low-carbon economy of the future will support many new jobs
and new companies alongside organizations that develop a clear vision of the future
and adapt to it. Automation of precision and repetitive tasks will continue to transfer
employment from the factory floor to those companies that design and manufacture machine tools of the future and those who specialize in artificial intelligence,
information technology, and data management and storage.

Decarbonizing the economy will not end all traditional jobs. Mining for coal
will decline, but mining for rare earth elements will grow in importance. With a
view to sustainability, recovery and recycling of these critical materials will gain in
importance. More workers will focus on ecodesign, the greening of the construction
sector and innovative ways to produce aluminum, steel, cement, and other essential
materials that are now made with carbon-intensive technologies.

The low-carbon strategy will feature sustainable buildings and smart cities,
oriented to urban well-being. The expansion of broadband Internet availability will
enable knowledge workers to relocate to rural areas and bring expanded opportunities to smaller communities. Consumer products with lower amounts of embodied
carbon, reduced plastic in content and packaging, and higher degrees of recyclability
will gain market share.

8.13   Stranded Assets

The term "stranded asset" describes the asset class of property, plant, and equipment that becomes obsolete before the end of its useful economic life. The transition to a low-carbon economy will create a substantial number of stranded assets as energy sources transition from fossil fuels to renewable sources. This will cause—and is already causing—companies to write down the value of these investments. Such write downs can cause companies to become less valuable and may lead to lower stock prices for publicly traded companies, to bankruptcy, or both. For example, in December 2020, the US oil major ExxonMobil announced that it would write down the value of $17–20 billion worth of natural gas assets in western Canada, the United States, and Argentina, and not develop them. ExxonMobil's share price declined in 2020 by about half, and the company reported financial losses in each of the first three calendar quarters of 2020, a stark reversal of fortune precipitated by cutbacks in oil demand during the coronavirus pandemic (Brower and McCormick 2020).

Stranded assets are not a new phenomenon. At the turn of the twentieth century, carriage makers and suppliers of buggy whips found their industries decimated by the arrival of self-propelled automobiles. When disruptive technologies arrive, new businesses flourish. Some businesses adapt and survive, others go out of business. The transition to a low-carbon economy will benefit some sectors while harming others. Investments are being made in energy storage systems that are needed to ensure the delivery of electricity when renewable energy sources are inadequate to meet demand (Dajani 2019). Faced with disruptive change, governments commit to prepare affected workers to perform the jobs of the future and to assist communities to adapt to newly ascendant economic models. Safety net policies include unemployment insurance, worker retraining, and financial assistance to companies adapting to new economic conditions.

At the beginning of the decade of the 2020s, the energy sector had already begun the process of transitioning away from fossil fuels to renewable energy. Wind and solar installations compete economically with fossil fuel plants. An increasing number of companies are committed to becoming "net-zero" carbon emitters by 2050. In 2018, New Zealand stopped issuing new permits for offshore oil exploration to meet its carbon neutrality goals. And in 2020, Denmark, the European Union's largest producer of oil, announced that it will not only stop issuing new offshore drilling permits but will also end the validity of existing drilling permits in 2050 (Farand 2020).

8.14   Inequality, Culture, and Consumerism

We should not overlook the importance of cultural attitudes toward consumption in the post-pandemic "new normal." Just as children raised by parents during the time of the Great Depression of the twentieth century were forever marked by exhortations to finish the food on their plates and turn off the lights when leaving a room, the post-pandemic generation may learn to value environmental protection more than gratification of immediate wants and needs. Some signs of this were present before the pandemic as Swedes coined the term "flygskam" (flight shame) in 2017 out of a conscious desire to reduce GHG emissions. It has led to increased train travel in Europe and to climate activist Greta Thunberg's voyage by sailboat to New York in 2019.

A large proportion of the world's population lives on far too little income to be presented with the choice of whether to fly or travel by train. Their individual carbon budgets are very modest, and their personal consumer choices do little to move the needle on the world's carbon emissions. It is not the same for the middle classes and wealthy in the world's developed countries, and the elites in the world's developing countries.

As Thomas Picketty observed, North America as a region stands apart for its high share of CO_2 emissions at more than 20% for 4% of the world's population. Moreover, it leads the world in carbon emissions inequality as measured by the number of individuals whose emissions exceed the world average of 6.2 tons per year. In his most recent book, *Capital and Ideology*, Piketty included a bar chart describing the global distribution of carbon emissions in the years 2010–2018. It showed that North America accounted for 36% of the world's share of individuals whose emissions are greater than 6.2 tons per year. It also accounts for 46% of the world's population whose emissions exceed the world average by 2.3 times, and 57% of the world's population that exceeds the world average by 9.1 times.

In contrast, Europe accounted for 20% of the world's population whose individuals exceeded the global average, China 15%, and the rest of the world approximately 29%. Piketty explains the high carbon intensity of the USA and Canada to consumer preferences for large homes, gas guzzling vehicles, and other lifestyle choices. He concludes from this data that "it would not be illogical for the United States to compensate the rest of the world for the damage it has done to global well-being" (Piketty 2020, p. 667).

Piketty does not mention North America's prominent role in oil and gas extraction and its continent-wide travel distances. It is well known that major oil-producing countries have high per capita CO_2 emissions, with Saudi Arabia at 18.48 tCO_2e, the USA at 16.56, Canada at 15.32, and Russia at 11.74 (Union of Concerned Scientists 2020). Moreover, less urban density in North America than is found in Europe makes public transportation a less convenient option and moving between all but the closest cities favors air travel over trains.

Piketty's point should not be completely discounted, however. North America, and the USA in particular, is responsible along with Europe for a disproportionate amount of historical emissions of CO_2 and other GHGs. These regions of the world benefited from early industrialization and achieved a high standard of living compared to the populations of less developed countries. These disparities resulted in the formulation of "common but differentiated responsibilities" for addressing climate change when the United Nations Framework Convention on Climate Change was signed, and the principle remains valid nearly thirty years later. No less a world figure than Pope

Francis framed in moral terms the excessive consumption and unsustainable use of the world's natural resources by the world's privileged inhabitants. In *Laudato Si'* the Pope called for preserving the world for future generations and working to ensure that the basic needs of all human beings are met (Pope Francis 2015).[4]

The challenge for governments, businesses, and consumers in a post-pandemic world that seeks to restore economic activity intentionally curtailed by shutdowns, working from home, and social distancing is "building back better" that addresses the twin crises of pandemic recovery and climate change. Solutions will be tested not only against decarbonization goals mandated by the Paris Agreement but also by the extent to which societies implement social safety nets to help dislocated workers and members of younger generations whose access to jobs and a career ladder were disrupted by the pandemic and its economic side effects.

Questions for Readers

1. What role should there be for nuclear energy in the low-carbon economy?
2. How should the housing and commercial building sectors be re-imagined to ensure that newly constructed buildings meet high energy-efficiency standards and climate resilience?
3. What role should green finance play in decarbonizing world economies?
4. What challenges confront manufacturers in transitioning to low-carbon products?
5. Can continuing business-as-usual lead the world to a net-zero future?

References

ADEME (2021) Tous les acteurs doivent agir collectivement pour la neutralité carbone, mais aucun acteur ne devrait se revendiquer neutre en carbone, April 2021 https://presse.ademe.fr/2021/04/avis-de-lademe-tous-les-acteurs-doivent-agir-collectivement-pour-la-neutralite-carbone-mais-aucun-acteur-ne-devrait-se-revendiquer-neutre-en-carbone.html. Accessed 7 May 2021

Ali K et al (2020) Issues, impacts, and mitigations of carbon dioxide emissions in the building sector. Sustainability 12:7427. https://doi.org/10.3390/su12187427.Accessed13June2021

Allenby B (1999) Industrial ecology. Prentice Hall

Bonafin J et al (2019) CO2 emissions from geothermal power plants: evaluation of technical solutions for CO2 reinjection in European Geothermal Congress 2019 Den Haag the Netherlands. https://europeangeothermalcongress.eu/wp-content/uploads/2019/07/291.pdf. Accessed 20 May 2021

---

[4] In paragraph 27 of the encyclical letter, the Pope writes: "Other indicators of the present situation have to do with the depletion of natural resources. We all know that it is not possible to sustain the present level of consumption in developed countries and wealthier sectors of society, where the habit of wasting and discarding has reached unprecedented levels. The exploitation of the planet has already exceeded acceptable limits and we still have not solved the problem of poverty."

BRE Group (2021) Encouraging positive social impact and equity using BREEAM. https://www.breeam.com/. Accessed 27 May 2021

Brower D, McCormick M (2020) ExxonMobil slashes capex and will write off up to $20bn in assets. https://www.ft.com/content/145765b3-2385-4d2f-a71b-82d9b81e85da. Accessed 15 January 2021

Carbon Brief (2019) Factcheck: how electric vehicles help to tackle climate change. https://www.carbonbrief.org/factcheck-how-electric-vehicles-help-to-tackle-climate-change. Accessed 5 June 2020

Carrington D (2020) Covid-19 lockdown will have 'negligible' impact on climate crisis—study. The Guardian, 7 August 2020. https://www.theguardian.com/environment/2020/aug/07/covid-19-lockdown-will-have-negligible-impact-on-climate-crisis-study. Accessed on 24 May 2021

Cochran KM, Townsend TG (2010) Estimating construction and demolition debris generation using a materials flow analysis approach. Waste Manage (30):2247–2254. Referred by Maderas, Cienc. Tecnol. vol.20 no.4 Concepción Oct. 2018. https://doi.org/10.4067/S0718-221X2018005041401. Accessed 30 March 2021

Commoner B (1971) The closing circle. Knopf, Alfred A

CRG Architects (2021) Bamboo skyscraper. https://www.arch2o.com/bamboo-skyscraper-crg-architects/. Accessed on 24 May 2021

Criqui P et al (2009) Les États et le carbone. https://www.researchgate.net/publication/47281181_Patrick_Criqui_Benoit_Faraco_Alain_Grandjean_2009_Les_Etats_et_le_carbone_Presses_Universitaires_de_France_France_192_p. Accessed 5 Feb 2021

Dajani M (2019) The remaining challenges in understanding energy storage as an investment, in greentechmedia. https://www.greentechmedia.com/articles/read/the-remaining-challenges-in-understanding-energy-storage-as-an-investment. Accessed 1 Apr 2021

Dunbar B (2013) https://www.nasa.gov/agency/sustainability/greenroofs.html. Accessed 30 Mar 2021

Dupin L (2021) Pour l'Agence internationale de l'énergie, tous les nouveaux projets pétroliers et gaziers sont désormais indésirables, in Novethic. https://www.novethic.fr/actualite/energie/transition-energetique/isr-rse/pour-l-agence-internationale-de-l-energie-tous-les-nouveaux-projets-petroliers-et-gaziers-sont-desormais-indesirables-149815.html. Accessed 18 May 2021

EEA (2019) Trends and projections in Europe 2019. https://www.eea.europa.eu/publications/trends-and-projections-in-europe-1. Accessed 16 Apr 2020

Ellen MacArthur Foundation (2015) Growth within: a circular economy vision for a competitive Europe. https://www.ellenmacarthurfoundation.org/assets/downloads/publications/EllenMacArthurFoundation_Growth-Within_July15.pdf. Accessed 6 June 2020

European Commission (2017) The chemical industry can achieve a 36% reduction in annual GHG emission in 2050, study shows. https://ec.europa.eu/jrc/en/news/chemical-industry-can-achieve-36-reduction-annual-greenhouse-gas-emissions-2050-study-shows. Accessed 28 Mar 2021

European Commission (2019) Development of a guidance document on best practices in the extractive waste management plans. https://op.europa.eu/en/publication-detail/-/publication/f18472f8-36aa-11e9-8d04-01aa75ed71a1/language-en/format-PDF/source-87989698. Accessed 6 June 2020

European Commission (2020a) Taxonomy report: technical annex. https://ec.europa.eu/info/sites/info/files/business_economy_euro/banking_and_finance/documents/200309-sustainable-finance-teg-final-report-taxonomy-annexes_en.pdf. Accessed 15 Apr 2020

European Commission (2020b) Sustainable finance: TEG final report on the EU taxonomy. https://knowledge4policy.ec.europa.eu/publication/sustainable-finance-teg-final-report-eu-taxonomy_en. Accessed 15 Apr 2020

Evans S (2018) In-depth: is Shell's new climate scenario as 'radical' as it says? https://www.eco-business.com/news/in-depth-is-shells-new-climate-scenario-as-radical-as-it-says/. Accessed 15 Nov 2020

Evan (2016) What Building Material (wood, steel, concrete) has the smallest overall environment impact?—debating science. https://blogs.umass.edu/natsci397a-eross/what-building-mat
erial-wood-steel-concrete-has-the-smallest-overall-environment-impact/. Accessed 28 Mar 2021

Farand C (2020) Denmark to phase out oil and gas production by 2050 in "watershed"
decision. https://www.climatechangenews.com/2020/12/04/denmark-phase-oil-gas-production-
2050-watershed-decision/. Accessed 10 Dec 2020

Fairley P (2009) Using CO2 to extract geothermal energy. MIT Technology review. https://www.tec
hnologyreview.com/2009/11/16/208159/using-co2-to-extract-geothermal-energy/. Accessed 27
Mar 2021

Foroudi L (2021) The tyranny of concrete. Financial Times 22–23 May 2021, pp 10–11

Frankfurt School of Finance and Management GmbH (2018) Global trends in renewable energy
investment 2018. Frankfurt School-UNEP Centre (Frankfurt am Main). https://europa.eu/cap
acity4dev/unep/documents/global-trends-renewable-energy-investment-2018. Accessed 27 Mar
2021

Gates B (2021) How to avoid a climate disaster: the solutions we have and the breakthroughs we
need. Knopf, Alfred A

Gerrard, Dernbach (2018) Legal pathways to deep decarbonization in the United States. Summary &
key recommendations. Environmental Law Institute

Gupta R (2019) Renewable energy holds huge potential to replace fossil fuels. Here is a breakdown of
the industry. https://via.news/analysis/renewable-replace-fossil-fuel-industry/. Accessed 27 Mar
2021

Hammond G (2021) The developers still betting on the London office market. Financial Times 17
May 2021, p 17

Hetzel J (2003) Haute qualité environementale du cadre bati enjeux et pratiques. AFNOR 2003

Hicks R (2020) 200 of world's largest corporations commit to net zero emissions by 2050, reverse
biodiversity loss and fight inequality in eco-business. https://www.eco-business.com/news/200-
of-worlds-largest-corporations-commit-to-net-zero-emissions-by-2050-reverse-biodiversity-
loss-and-fight-inequality/. Accessed 30 March 2021

Hodgson C (2021) Arctic data stoke global warming fears. Financial Times, 28 May 2021, p 2

Hook L, Hodgson C (2021) G7 countries agree to end international funding of coal. Financial Times,
22–23 May 2021, p 4

Hsu SL (2010) Life cycle assessment of materials and construction in commercial structures: variability and limitations. Massachusetts Institute of Technology, June. https://dspace.mit.edu/han
dle/1721.1/60767. Accessed 30 Mar 2021

IEA (2018) Energy efficiency report 2018. Analysis and outlooks to 2040. October 2018. https://
www.iea.org/reports/energy-efficiency-2018. Accessed 6 Mar 2021

IEA (2021) Net Zero by 2050: a roadmap for the global energy sector. https://iea.blob.core.win
dows.net/assets/ad0d4830-bd7e-47b6-838c-40d115733c13/NetZeroby2050-ARoadmapfortheG
lobalEnergySector.pdf. Accessed 18 May 2021

IPCC (2018) Masson-Delmotte et al. Global Warming of 1.5 °C. An IPCC special report on the
impacts of global warming of 1.5 °C above pre-industrial levels and related global greenhouse
gas emission pathways, in the context of strengthening the global response to the threat of climate
change, sustainable development, and efforts to eradicate poverty. https://www.ipcc.ch/site/ass
ets/uploads/sites/2/2019/06/SR15_Full_Report_Low_Res.pdf. Accessed 31 May 2021

IIF (2020) Taskforce on scaling voluntary carbon markets. Institute of International Finance https://
www.iif.com/Portals/1/Files/TSVCM_Consultation_Document.pdf. Accessed 4 Dec 2020

IRENA (2020) Green hydrogen cost reduction scaling up electrolysers to meet 1.5 °C
climate goals. https://irena.org/-/media/Files/IRENA/Agency/Publication/2020/Dec/IRENA_
Green_hydrogen_cost_2020.pdf. Accessed 27 Mar 2021

ISO (2006) ISO 14044 environmental management—life cycle assessment—requirements and
guidelines. https://www.iso.org/standard/38498.html. Accessed 13 June 2021

ISO (2018a) ISO 14067 greenhouse gases—carbon footprint of products—requirements and
guidelines for quantification. https://www.iso.org/standard/71206.html. Accessed 13 June 2021

ISO (2018b) ISO 14064-1:2018 Greenhouse gases—part 1: specification with guidance at the organization level for quantification and reporting of greenhouse gas emissions and removals. https://www.iso.org/obp/ui/#iso:std:iso:14064:-1:ed-2:v1:en. Accessed 18 Jan 2021

ISO (2015a) ISO 14001 environmental management systems—requirements with guidance for use. https://www.iso.org/standard/60857.html. Accessed 13 June 2021

ISO (2015b) ISO 9001 quality management systems—requirements. https://www.iso.org/standard/62085.html. Accessed 13 June 2021

ISO (2021) ISO 14021 AMD 1. https://www.iso.org/standard/81242.html. Accessed 27 May 2021

Jidb175 (2018) Circular business model. https://commons.wikimedia.org/w/index.php?curid=686 08849. Accessed 4 Dec 2020

Johnston R (2018) Ecosystem services. Encyclopedia Britannica, 5 Jan 2018. https://www.britan nica.com/science/ecosystem-services. Accessed 30 Mar 2021

Maykuth A (2019) Are these tiny, 'inherently safe' nuclear reactors the path to a carbon-free future? Phys.org 22 March 2019, ©2019 Philly.com, distributed by Tribune Content Agency, LLC. https://phys.org/news/2019-03-tiny-inherently-safe-nuclear-reactors.html. Accessed 17 May 2021

Milne R (2020) Sweden's LKAB plans $47bn push into carbon-free iron ore. https://www.ft.com/content/0cbc20c6-d781-44da-a77b-cb7679f9820a. Accessed 15 Feb 2021

Petrova M (2020) Powering the future. Green hydrogen is gaining traction, but still has massive hurdles to overcome. CNBC 4 December 2020. https://www.cnbc.com/2020/12/04/green-hyd rogen-is-gaining-traction-but-it-must-overcome-big-hurdles.html. Accessed 27 March 2021

Piketty T (2020) Capital and ideology. Belknap Press of Harvard University Press

Francis P (2015) On care for our common home. http://www.vatican.va/content/dam/francesco/pdf/encyclicals/documents/papa-francesco_20150524_enciclica-laudato-si_en.pdf. Accessed 19 Feb 2021

Prescient & Strategic intelligence (2020) Green chemicals market overview. https://www.psmark etresearch.com/market-analysis/green-chemicals-market-outlook. Accessed 28 March 2021

Raval A (2021) Shell's defeat spells trouble for other polluters. Financial Times, 28 May 2021, p 9

Richter A (2019) Geothermal hot springs used in Roman times could heat Bath Abbey, England. Work is underway to utilize Roman Bath's hot geothermal water to heat Bath Abbey, a former monastery in Somerset, England. 30 January 2019. https://www.thinkgeoenergy.com/geothermal-hot-springs-used-in-roman-times-could-heat-bath-abbey-england/. Accessed 27 Mar 2021

REN 21 (2020) Key findings of the renewables 2020 global status report. https://www.ren21.net/reports/global-status-report/. Accessed 28 Mar 2021

Science Based Targets Initiative (2021) https://sciencebasedtargets.org/about-us. Accessed 22 May 2021

Stadler I, Sterner M (2018) Urban energy storage and sector coupling. In: Urban energy transition, 2nd edn. https://www.sciencedirect.com/topics/engineering/chemical-energy-storage. Accessed 28 Mar 2021

Stahel R (1982) "The product-life factor" submitted to the Mitchell prize competition on sustainable societies in Houston. TX, USA

Turnbull M (2021) Hydro storage can help fuel the clean energy transition. Financial Times, 9 March 2021, p 17

Turner G, Grocott H (2020) The global voluntary carbon market. Dealing with the problem of historic credits. 4 December 2020. Trove Research https://trove-research.com/research-and-insight/the-global-voluntary-carbon-market-dealing-with-the-problem-of-historic-credits-dec-2020/. Accessed 27 May 2021

UNFCCC (2021) Total potential supply of CERs from end KP 1st CP to 2020. https://cdm.unfccc.int/Statistics/Public/files/202102/CER_potential.pdf. Accessed 12 Mar 2021

Unger N (2014) Yale study shows how conversion of forests to cropland affected climate https://environment.yale.edu/news/article/yale-study-looks-at-how-global-conversion-of-forests-to-cropland-affected-climate/. Accessed 30 Mar 2021

Union of Concerned Scientists (2020) Each country's share of CO2 emissions, Updated 12 August 2020. https://www.ucsusa.org/resources/each-countrys-share-co2-emissions. Accessed 22 May 2021

US EPA (2020) Fast Facts U.S. transportation sector greenhouse gas emissions 1990–2018. https://nepis.epa.gov/Exe/ZyPDF.cgi?Dockey=P100ZK4P.pdf. Accessed 5 June 2020

US GBC (2021) U.S. Green Building Council. https://www.usgbc.org/. Accessed 27 May 2021

World Bank (2016) Greenhouse gases from geothermal power production. Technical report 009/16   http://documents1.worldbank.org/curated/en/550871468184785413/pdf/106570-ESM-P130625-PUBLIC.pdf. Accessed 27 Mar 2021

World Bank (2021) Global gas flaring reduction partnership (GGFR). https://www.worldbank.org/en/programs/gasflaringreduction. Accessed 27 May 2021

Chapter 9
Conclusions

9.1 Assessing Climate Policy

In 2019, the Intergovernmental Panel on Climate Change (IPCC) issued a special report on oceans and the cryosphere (frozen lands) that raised urgent concerns about ocean acidification and the impacts of warming on the cryosphere. It emphasized that choices made now were critical for minimizing impacts from melting ice caps and glaciers, thawing of permafrost, sea-level rise, and increasing intensity of tropical cyclones. It documented the consequences of these impacts on marine species and the humans who rely upon them (IPCC 2019).

It is fair to conclude that, during the first two decades of the twenty-first century, the world has fallen short of achieving the amount of emission reductions needed to keep global warming to a range of 1.5–2 °C. Some policies have worked better than others, but policies have also faced the headwinds of increased population and economic growth. Until 2020, Kyoto mechanisms applied only in a limited number of economies, with the two largest emitting nations—China and the USA—remaining on the sidelines.

The Paris Agreement rectified the problem of less than universal commitment to action, but critics quickly pointed out that the level of ambition communicated by countries' first nationally determined contributions (NDCs) was insufficient to achieve the goals of the agreement by 2050. In the first two decades of the twenty-first century, world emissions of carbon dioxide (CO_2) and other greenhouse gases (GHGs) continued to rise. Increases were driven particularly by developing countries in Asia including China and India. Figure 9.1 below shows both that China exports the most embodied CO_2 in products and that the largest share of those goods arrives in the United States. China is the world's largest emitter of GHGs. If China's exported CO_2 were excluded from its national inventory, the amount of CO_2 China was responsible for in 2014 would decline by 13% (Hausfather 2017).

Climate policies in the EU and some jurisdictions in North America performed better with respect to emissions growth, but absolute emission rates remain relatively high on a per capita basis, particularly in the USA (see Fig. 9.2).

J. C. Shideler and J. Hetzel, *Introduction to Climate Change Management*,
Springer Climate, https://doi.org/10.1007/978-3-030-87918-1_9

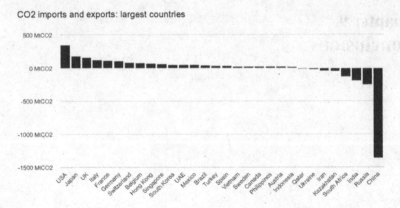

Fig. 9.1 CO_2 imports and exports: largest countries (Hausfather 2017)

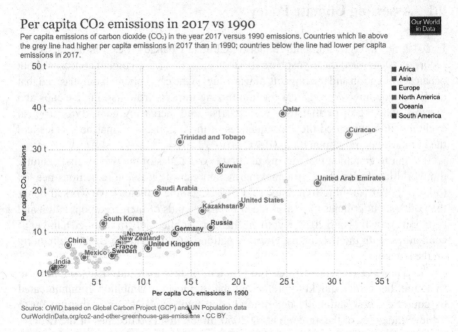

Fig. 9.2 Per capita CO_2 emissions in 2017 compared to 1990 (Our World in Data 2020)

The USA has just 4% of the world's population but was responsible in 2015 for about 16% of the world's GHG emissions (see Fig. 9.3).

The numbers reflected in Fig. 9.3 do not include the amount of embodied CO_2e emissions in imported goods. The USA in 2014 imported goods valued at approximately $2.3 trillion. Imported trade goods include those grown, produced, or manufactured in other countries, including those of domestic manufacture which have been worked on abroad before reentering the USA (Statista 2021). According to figures

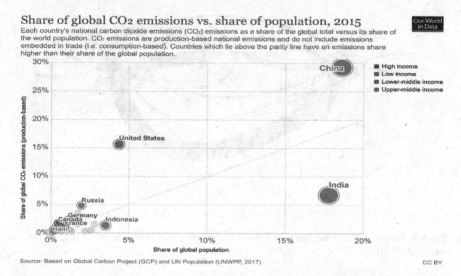

Share of global CO₂ emissions vs. share of population, 2015

Each country's national carbon dioxide emissions (CO₂) emissions as a share of the global total versus its share of the world population. CO₂ emissions are production-based national emissions and do not include emissions embedded in trade (i.e. consumption-based). Countries which lie above the parity line have an emissions share higher than their share of the global population.

Source: Based on Global Carbon Project (GCP) and UN Population (UNWPP, 2017)   CC BY

Fig. 9.3 Share of global CO_2 emissions versus share of populations (Our World in Data 2015)

published by CarbonBrief, the embodied CO_2 in trade goods imported into the USA in 2014 amounted to 352 Mt which was equivalent to 6% of US domestic emissions. Over the period 1990–2014, the amount of embodied CO_2 in imported trade goods rose from 9 to 17% of US national CO_2 emissions (Hausfather 2017).

On a per capita basis, the USA has by far the highest emissions of CO_2 than any other country except those whose economies are dominated by the production of oil and gas (see Fig. 9.2). Economic growth and production for export are two reasons why China's rate of emissions has increased substantially since the signing of the Kyoto Protocol. Measured in Gross Domestic Product (GDP), China's annual growth rate averaged 9.23% from 1989 until 2020 (Trading Economics 2021). China is also the world's most populous nation whose per capita income rose more than tenfold from 2000 to 2019 (China Power 2017). World flows of CO_2 emissions are not limited to the trading relationship of China and the USA, as illustrated in Fig. 9.4.

Given the world's rising population and desire among many developing countries to "catch up" to the standard of living enjoyed in the USA and Europe, it is difficult to see how global emissions of GHGs will decrease on an absolute basis between 2021 and 2050 without a significant increase in the level of ambition to reduce emissions by the world's richest and most industrialized economies. Despite the example shown by the state of California, current policy initiatives are not likely to move the needle far enough to achieve the goals set out in the Paris Agreement. Even assuming a nationwide US climate policy equivalent to that pursued in California, the world will need a redoubling of effort in the world's highest emitting economies and the adoption of sustainable climate policies by countries in other regions around the world.

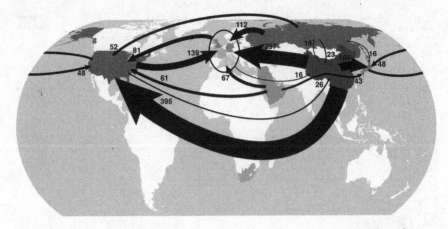

Fig. 9.4 Largest interregional fluxes in 2004 of emissions embodied in trade (Mt CO_2 y^{-1}) (Davis and Caldeira 2010)

9.2   The Carbon Markets

As discussed in Chap. 2, the design and implementation of carbon markets, particularly with respect to cap-and-trade policies implemented in the European Union and in North America, have provided industry with options for meeting GHG emission reduction targets. As participants ourselves in the infrastructure of these markets—until 2021 John Shideler co-managed a group of validators and verifiers for NSF Certification, LLC, whose number included Jean Hetzel—we believe that verified offset credits issued by the major regulatory and voluntary programs can represent real, permanent, accurate, verifiable, and enforceable emission reductions. That is not to say, however, that there are not issues associated with the use of this policy mechanism that need to be addressed. A summary of the issues identified in Chap. 2 includes the following:

– The price of offset credits has been too low to properly incentivize industry to decarbonize its own processes and products.
– The additionality of some offset credits may legitimately be questioned.
– Regulatory cap-and-trade programs do not cover all emissions, including those associated with imported goods that cross borders without a carbon tax.

Controversies over the effectiveness of employing offset credits to compensate for current GHG emissions are not new. Writing in 2009, Augustin Fragnière titled a book on this subject "La compensation carbone: illusion ou solution?" ("Carbon offset credits: Illusion or solution?") (Fragnière 2009). Fragnière wrote too soon to accurately describe issues related to Joint Implementation projects under the Kyoto Protocol. Instead, he focused his attention on Clean Development Mechanism (CDM) and voluntary market projects. He faulted the voluntary market for operating outside a compulsory regulatory framework, for uneven quality standards compared to the

CDM, for the absence of a centralized institution for control, and above all for the theoretically unlimited use of its instruments (Fragnière 2009). Nothing, he said, prevents an organization from freely compensating the entirety of its emissions on the voluntary market. "This is the origin of a key concept in the problem of compensation: carbon neutrality" (Fragnière 2009, pp. 36–37).

Declarations of intent to achieve carbon neutrality expanded in 2020 among multinational organizations in 2020 (Nguyen 2020). Increased commitments to combat the crisis of climate change across all fronts are welcome and urgently needed, whether by national or subnational governments or by "non-state actors."[1] As attractive as a concept carbon neutrality may be, achieving it is difficult. One way to achieve carbon neutrality is to revamp the operations of an enterprise, so that its total emissions and removals equal "net zero." This is far easier said than done, as the technical means to achieve net zero currently do not exist for most organizations. So, declarations of carbon neutrality depend upon the use of offsetting carbon credits to bring an organization's GHG emission and removal accounts into balance.

We do not challenge the concept of "carbon neutrality," but only its definition. Indeed, carbon neutrality represents the "Holy Grail" of climate mitigation efforts: the ability of the world's economies to reduce to zero anthropomorphic loading of the atmosphere with additional CO_2 and other GHGs and to end humankind's disruption of the natural carbon cycle (see Chap. 8, "The Path to Net Zero"). The problem, as Fragnière and others have pointed out, is that not all carbon offset credits can be demonstrated to meet the strictest standards for quality or pass the tests of additionality and permanence. Moreover, the price signal offered in 2020–2021 in voluntary carbon credit markets and in unused CDM credits is woefully inadequate to incentivize "deep decarbonization." This situation has caused some commentators to equate today's carbon offset credit market with the "papal indulgences" sold by the Roman Catholic church in the fifteenth and early sixteenth centuries (Hook 2020).

As Fragnière observed in his 2009 book, the English word "offset," as defined by the Oxford English Dictionary (1990), means "a compensation, a consideration, or amount diminishing or neutralizing the effect of a contrary one" (Fragnière 2009). What it does not do is serve as an indication that the activities maintained by the organization making a claim of carbon neutrality will ever achieve the level of decarbonization needed to reduce global warming to no more than 2 °C by 2050. We return now to our discussion of how the first two decades of the twenty-first century have produced disappointing results when compared with the magnitude of the present climate crisis.

---

[1] The term "non-state actor" is commonly used by diplomats to refer to individuals, corporations, private financial institutions, and nongovernmental organizations of any type.

9.3   Europe's Experience with the Emissions Trading System

We presented in Chap. 2 the European Union's experience with its Emissions Trading System (ETS). We found that the price of carbon as reflected in European Union Allowances (EUAs) fell far short of the carbon price of $40–80 that the World Bank in 2019 estimated was necessary by 2020 to reach goals of the 2015 Paris Agreement (S&P global 2020a; b). Indeed, in 2020, EUAs traded mostly in a range around €25 per ton ($30). A year later (2021), EUA prices had doubled to €60 or more, a price on carbon that observers believed would finally incentivize efforts to decarbonize. The rise in EUA prices had taken many years during which time investment decisions were made based on an unrealistically low perception of the "cost of carbon."

The higher EUAs advanced in price during 2021, the louder European industry and policy makers called for a mechanism to tax goods imported into the European Union that had been produced in low- or no-carbon tax economies. The favored policy tool was a carbon border adjustment mechanism that would protect European industry from competing against steel and other products produced without similar costs of carbon (Marcu et al 2021).

9.3.1   Interim Emission Reduction Targets in the EU

To help fight climate change, EU leaders adopted in October 2014 the 2030 climate and energy framework, which includes binding targets to cut emissions in the EU by at least 40% below 1990 levels by 2030. The EU is launching various initiatives to reach these targets. One of them is the Effort Sharing Regulation which obliges European member states to achieve carbon emissions reductions in the twelve years from 2019 through 2030.

Projections of GHG emission trends for reductions suggests that the EU will have difficulty meeting its 2030 target with current member state commitments (see Fig. 9.5). The task of reducing emissions in Germany was made more difficult by the country's decision, after the Fukushima nuclear power plant accident in 2011, to close its operating nuclear reactors. The decision resulted in Germany reopening coal-fired power plants that generate electricity at far higher GHG emission rates. France, on the other hand, lagged in the development of renewable energy sources due to the dominance of nuclear power in its national grid supply.

9.3.2   The European Green Deal

The "European Green Deal" was launched in 2019 by the new European Commission as a roadmap for making the EU's economy sustainable by turning climate and

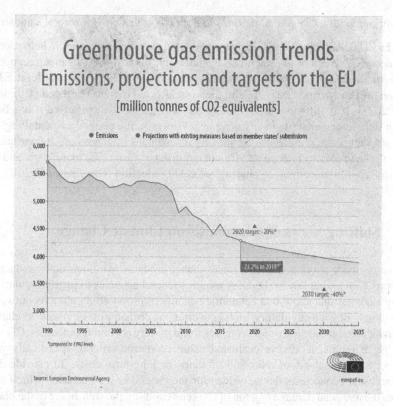

Fig. 9.5 GHG emissions compared with the 2020 and 2030 targets (EU Parliament 2018)

environmental challenges into opportunities across all policy areas and making the transition just and inclusive for all (European Commission 2019). The objectives of the European Green Deal are to:

– boost the efficient use of resources by moving to a clean, circular economy
– restore biodiversity and cut pollution
– encourage needed investments and financing for a just and inclusive transition to a low-carbon economy.

The EU aims to be climate neutral in 2050. To do this, the European Council proposed a European Climate Law turning political commitments into a legal framework and stimulating investment. Achieving EU targets will require action by all sectors of the economy, including:

– investing in environmentally beneficial technologies
– supporting industry innovation
– deploying cleaner, cheaper and healthier forms of private and public transport
– decarbonizing the energy sector
– ensuring buildings are more energy efficient

– working with international partners to improve global environmental standards.

The EU will also provide financial support and technical assistance to help people, businesses, and regions that are most affected by the move toward the green economy. This is called the Just Transition Mechanism and will help mobilize at least €100 billion over the period 2021–2027 in the most affected regions (European Commission 2021). It is difficult to pass judgment on a global policy before it has been given time to demonstrate its effectiveness. However, the nongovernmental organization Greenpeace criticized it for making investments in natural gas infrastructure in central and eastern Europe and for not requiring governments like Poland and the Czech Republic to commit to phasing out coal production (Greenpeace 2020).

9.4   Shifting Views of Economists on Climate Change

Decisions made about acting on climate change are heavily influenced by economic analysis. Governing law in the USA requires federal agencies to justify regulations using analyses that show that economic benefits of proposed regulations outweigh their costs. Beyond regulatory tools, government decision makers seek to change behavior through subsidies and tax policies that modify economic outcomes. Climate change, it seems, is always evaluated using economic criteria, specifically with respect to the analysis of risks and of short- or long-term costs. To provide data for analysis, economists devise values for the "social cost of carbon" which estimates the marginal damage of a ton of CO_2 emissions on the well-being of the planet and its inhabitants.

A study titled "Gauging Economic Consensus on Climate Change" published in March 2021 reported on the results of a questionnaire sent to 3000 economists who had published articles on climate change in economic journals. The study, which followed up a similar poll conducted by the same authors in 2015, found that respondents' views on climate change had evolved in the intervening years. Survey results from 2015 are provided in brackets after the 2021 answers to the same question (Howard and Sylvan 2021, p. 12):

"Which of the following best describes your views about climate change?

1.   Immediate and drastic action is necessary 74% (50%).
2.   Some action should be taken now 24% (43%).
3.   More research is needed before action is taken 2% (5%).
4.   This is not a serious problem 1% (1%).
5.   No response (1%)".

The recognition by nearly three-quarters of respondents that immediate and drastic climate change action is necessary represents a notable shift in urgency from results tallied only five years previously. The poll also demonstrates that skeptics constitute an extremely small minority.

The second result of this study highlighted why economists are most concerned by climate change (Howard and Sylvan 2021, p. 13):

1. 52% observed extreme weather events attributed to climate change
2. 31% noted new findings in climate science
3. 29% noted new findings in climate economics and the social sciences.

The numbers reveal that some economists cited more than one reason for explaining changes in their concerns about climate change. This can be expected as scientific evidence mounts for the urgency of acting on climate change and the evidence is available for all with an open mind to see.

A different question in the survey asked about the link between climate change and growth: "What is the likelihood that climate change will have a long-term, negative impact on the growth rate of the global economy?" "Extremely likely," answered 40% of the economists. "Likely" was the choice of 36%. These answers correlate closely with the economists' response to the first question about the need for immediate and drastic action (Howard and Sylvan 2021, p. 15).

Economists also were persuaded that climate change increases inequality among countries and within countries, as 89% of respondents said that such results were extremely likely or likely (Howard and Sylvan 2021, p. 15).

Another question asked about potential costs and benefits: "Many government entities have set goals to reach net-zero GHG emissions by roughly mid-century (this would be consistent with a global average surface temperature limit of 1.5°–2 °C according to many projections). Are the expected benefits of mid-century net-zero GHG targets likely to outweigh the expected costs? Please account for any relevant co-benefits and co-costs in your implicit present-value estimates." Thirty-one percent of economists answered "extremely likely" while 36% opined "likely." This suggests that 66% of respondents support pursuing climate change policies that they believe will offer clear benefits for the economy (Howard and Sylvan 2021, p. 28).

The last question introduces our next paragraph: "Given the unprecedented events of 2020, please estimate the % change in global GDP and global greenhouse gas emissions (CO_2e) from 2019 to 2020, without using outside sources like the internet." The economists median estimate for global GDP change from 2019 to 2020 was − 3%. Their median estimate for changes from 2019 to 2020 in global GHG emissions was −3.5% (Howard and Sylvan 2021, p. 28).

9.5 Transition Risks and Financial Risks

The vision expressed in this book is for a convergence of commitments to address climate change by governments, industry, and the nongovernmental sectors, supported by majorities of citizens in countries around the world. Governments have a primary role to play in setting policies and adopting frameworks that will incentivize the private sector to decarbonize. Industry bears the primary responsibility to innovate technical solutions and to oversee their implementation, supported by appropriate government policies where needed. Nongovernmental organizations including standards setting bodies, multilateral organizations, and advocacy groups will help reduce friction during the transition by identifying roadblocks and inequities, encouraging transparency, and devising solutions.

9.5.1 Government Policies

We have highlighted efforts by governments to reduce greenhouse gas emissions by the adoption of carbon trading mechanisms such as the EU ETS and the California cap-and-trade system. Both Emissions Trading Systems have contributed to reductions of greenhouse gas emissions and the development of more energy-efficient industries in their jurisdictions. In California's case, the cap-and-trade system was one of many initiatives identified in the Global Warming Solutions Act of 2006 to address climate change. Some jurisdictions, such as British Columbia and Sweden, have imposed carbon taxes. The results have been helpful but insufficient. Policy frameworks have encountered the headwinds of growing populations, and results achieved to date have not put the world on a path to economy-wide decarbonization that will limit global warming to between 1.5 and 2 °C by 2050.

We do not presume to present optimal technical solutions for each sector of the economy, or blueprints for how to decarbonize every sector. Our study of existing climate change management strategies does lead us to conclude that governments must play a more active role in guiding the transition. They can do this by adopting regulations requiring the abatement of carbon pollution and levying taxes on GHG emissions. Governments also have a role to play in providing safety nets for workers in industries that need to disappear and in supporting the education and training of the workforce needed for tomorrow's economy.

Securities regulators can require more climate disclosure by publicly traded companies, in line with new and emerging international standards. Financial regulators such as central banks and supervisory agencies can require financial institutions to assess borrowers' physical climate risks and transition risks, and set limits on the amount of climate risk-related exposure that regulated financial institutions are allowed to maintain.

9.5.2 Industry Action

Early-stage climate action has made industry more energy efficient. This progress is to be applauded, but it is not enough. Hard-to-abate sectors such as cement, steel and aviation need government support through the adoption of regulations that will make alternative approaches economically competitive with traditional practices. "Green" cement and direct reduced iron are promising technologies (see Chap. 8). The bioeconomy offers the aviation industry access to sustainable alternative fuels (SAF) that are functionally equivalent to petroleum-based jet fuel. Already leading airlines and cargo operators are working with SAF producers to increase commercialization of this low-carbon option. During the transition to a low-carbon economy, however, policy support is needed to bridge the price gap between conventional jet fuel and SAF.

9.5.3 Nongovernmental Organizations

For these concluding paragraphs, we expand the meaning of "nongovernmental organizations" to include not just traditional advocacy groups often referred to by this name, but also standard-setting bodies, multilateral organizations such as the G20, and other parties that do not fit in the categories of "government" or "industry." Such organizations play an important role in defining the terms and norms of behavior that will be used to manage the transition, influencing policy makers, and informing the debates that will accompany the transition to a low-carbon economy.

As standards writers ourselves, we acknowledge the influence that standards have and their potential to effect change. The basis for "counting carbon" discussed in Chap. 3 stems from international standards ISO 14064 Part 1 and ISO 14067, as well as the GHG Protocol standards published by the World Resources Institute and the World Business Council on Sustainable Development. Counting carbon generates data to inform decision making on both micro- and macro-economic levels.

As a result of the Great Recession of 2008–2009, the group of 20 leading economies (commonly known as the "G20"), established a Financial Stability Board (FSB) to promote the reform of international financial regulation and supervision. The FSB has since recognized the significance to the international financial system of both physical and transition risks associated with climate change. The FSB's Task Force on Climate-Related Financial Disclosures provides a blueprint for companies to assess their climate risks and make decisions based on scenarios of trends in climate change and transition policies. The FSB has a continuing role to play in establishing frameworks for managing climate risks and the transition to a low-carbon economy.

9.6 Lessons from the Pandemic of 2020

Before developing the last paragraphs, we mourn the loss of the millions of persons around the world who died from covid-19 and extend our condolences to their families.

The economic crisis caused by covid-19 lockdowns appears to be deeper than previous recessions. Some economic experts attributed to it a loss of 5% of GDP. The crisis is not a monetary one, unlike 2008. Banks have cash and do not need more liquidity. A key pandemic response in developed Western economies has been extending financial support to businesses and individuals impacted by locking down

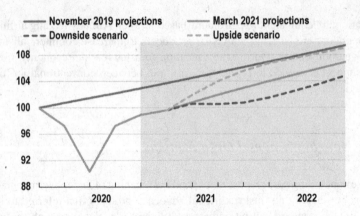

Fig. 9.6 Shock of the covid-19 pandemic on worldwide growth (OECD 2021)

some economic activity and encouraging all but nonessential workers to stay home. The resort to isolation was similar to Europeans' response to the spread of death and disease during the Black Death (1342–1353). In our modern pandemic, economic sectors were unevenly impacted. The travel, hospitality, and restaurant sectors were among the first to suffer. Office buildings emptied and international trade declined.

Figure 9.6 chronicles the immediate impact on economic activity in 2020 caused by the pandemic, but by the end of 2020 the net decline in US economic activity was only −2.9%. The situation cannot be compared with 1929 crisis which reduced US economic activity by 29.2% (1929–1933). Instead, the economy exhibited remarkable resilience and certain sectors such as home improvement and computer products increased sales during 2020. The new US administration of President Joe Biden increased expansionary policies to stimulate the economy and return as soon as possible to the previous economic growth trends.

Economists in 2020 debated the shape of the economic recovery. After a dramatic sell-off in stock prices in March 2020, some experts predicted a V-shaped recovery. Data seemed to support a powerful bounce back but many observers suggested the shape of the recover was more like a "K," with some sectors and individuals benefiting from rising economic activity while others risking permanent economic damage or decline.

The 2020 pandemic caused worldwide shocks. More than a year after the World Health Organization officially designated covid-19 as a pandemic, international borders remained mostly closed to travelers in many parts of the world. But the effects remained uneven. China returned to growth in 2021 after a period of intense lockdowns, and the USA and the United Kingdom sensed increased optimism after successfully rolling out vaccination campaigns. Meanwhile poor countries in South America and Africa waited for vaccine supplies to reach them.

9.6.1 Public Health Care

In the first two decades of the twentieth century, many governments focused on reducing the costs of public health services. Around the world, health expenses in 2014 accounted for 9.9% of GDP, but the public health system received only 6% of that financing. The most unequal health system was in the USA where health expenses represented 17.1% of GDP but only 8.2% was allocated to public health delivery (DNV GL 2015). This imbalance in resource allocation in the USA may explain why poor people were disproportionately impacted by illness and death during the covid-19 pandemic.

Healthcare systems in Europe support public health to a much greater extent with the exception of the United Kingdom whose National Health Service strained to meet public needs with a more limited allocation of resources. Health expenses in the United Kingdom represented only 9.1% of GDP, less than the world average of 9.9%, and two points lower than France at 11%.

The pandemic has shown the importance of delivering public health equitably. According to some health economists, a good level of protection can be provided to people for expenditures that average around 14% of GDP. Public health involves more than the ability to see doctors and be admitted to hospitals. It includes access to safe drinking water, operation of sanitary sewage systems, and availability of adequate housing.

Many countries are experiencing infrastructure gaps with the need for increased investment in public infrastructure, medical equipment, and software. Artificial intelligence used in the delivery of medical services is growing as technology can help identify the first outbreaks of illnesses. Investments in public health services can help prepare nations and communities to respond to future pandemics. The investments could be financed by long-term loans or bonds with state guaranties and could stimulate innovation by IT companies. 3D printing can bring down the cost of medical equipment and support health care in developing countries.

9.6.2 Behavioral and Life Style Changes

Some individual and collective behaviors are likely to be permanently modified because of the pandemic. Working from home is likely to remain popular among workers who have proven their ability to maintain productivity using the Internet and communications technologies. Technology platform companies such as Zoom were big beneficiaries of pandemic population lockdowns. Distributed work teams may recede from their pandemic heights due to corporate managers' desires to bring their workforces back to centralized offices. They are likely to be only partially successful. Recognizing this, some chief financial officers plan to let some office space leases expire and downsize their working spaces to accommodate more decentralized work forces. As a consequence, office building occupancy, especially in lower graded office buildings, is likely to decline. Average office rents may decline in large cities such as New York, London, Frankfurt, Singapore, or Tokyo.

Some owners of commercial real estate may face higher risks of insolvency if the current trend of low workforce density during working hours in rented spaces continues. Conversions of some office buildings to apartments may ease housing shortages that are widespread throughout the world. A trade-off from such conversions is that local authorities will then need to make investments in public services such as schools, parks, and hospitals if cities become more densely populated.

In some parts of the world, the pandemic has increased demand for fresh food and locally grown fruits and vegetables. Organic farms with local distribution to city dwellers could raise the quality of life for many people. Public markets could make use of underutilized car parks. It takes on average five years to convert a farm to organic production. During this time, the farmer's revenue is diminished due to the transition from conventional to organic food production and the misalignment with previous commercial sales relationships. Finance can assist farmers with the conversion by offering low interest rate loans.

9.6.3 Economic Impacts

In January 2020, when China put in place hard confinement, European and US economic activities were operating as normal. Some companies that have Chinese supply chain partners for automobiles or pharmaceuticals were looking at building inventory and seeking other solutions to maintain two weeks of supply. Confinement of residents in Wuhan was lifted after ten weeks, thousands of deaths, and slowed or interrupted industrial production. At about that time, the World Health Organization declared a worldwide pandemic and, in some sectors, economic activities in Europe and the USA slowed to a crawl or ground to a halt.

Aviation was a hard-hit sector. Some weak airlines sought protection in bankruptcy and other turned to their governments for assistance. In Europe, Air France and Lufthansa were practically nationalized. Tourism declined precipitously as countries closed their borders in an effort to stop the spread of covid-19 and its mutations. Cruise liners remained sidelined after passengers on their last sailings fell ill to the virus.

The rapid development and deployment in the first half of 2021 of covid-19 vaccines raised hopes that at least a part of the Northern Hemisphere summer travel season of 2021 could be salvaged. Consumers fortunate enough to remain employed during the pandemic had increased their savings and were eager to "return to normal" at the earliest opportunity. Still, the recovery of tourism to 2019 levels was unlikely to occur until at least 2022 if not 2023 due to ongoing health restrictions. Sectors affected include transportation, hotels, and restaurants. Ironically, prices for rental cars increased significantly in the USA in mid-2021 because car rental companies had sold off many of the idled vehicles in their fleets in 2020 and were short of product to rent when demand began to pick up in the second quarter of 2021.

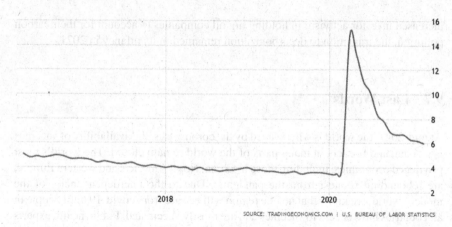

Fig. 9.7 US unemployment rate over five years using data from the US Bureau of labor statistics (Trading Economics.com 2021)

The pandemic's economic impacts fell largely on small businesses who lacked the balance sheets to finance months or years of stalled or low levels of business activity. Most governments in North America and Europe offered new financing to help businesses survive the pandemic and to reinforce their social safety nets for displaced workers. Pandemic aid helped many but some businesses were lost. Women in particular left the workforce to care at home for younger children no longer able to attend school. Unemployment fell disproportionately on minority and less well-educated workers.

Data from the US Bureau of Labor Statistics show the extent of the shock to employment in 2020 and its rapid return toward normal. Figure 9.7 tracks rates of unemployment which shot up in late March 2020 to reach a peak of 15.9% before gradually declining back to around 6%. This compares to a US unemployment rate before the start of the pandemic of 3.8%.

Mass unemployment of the kind recorded in the USA was largely avoided in western and northern Europe where social safety nets cushioned the effects of the pandemic on unemployment.

The skies cleared over many polluted cities during the pandemic as demand for transportation fuel fell dramatically during lockdowns. According to an International Energy Agency (IEA) report, "the most severe plunge in energy demand since the second war would trigger multidecade lows for the world's consumption of oil, gas while renewable energy continued to grow. The steady rise of renewable energy combined with the collapse in demand for fossil fuels means clean electricity will play its largest ever role in the global energy system this year (2020) and help erase a decade's growth of global carbon emissions" (Ambrose 2020). The prediction was premature as rebounding economic activity in 2021 made clear that only increased ambition for emission reductions, higher costs on carbon pollution, and continued policy interventions would place world economies on a trajectory to net zero by 2050. Despite higher than ever corporate commitments to join the "race to net zero" and increased investor activism in holding big oil companies to account for their carbon pollution, the transition to decarbonization remained in its infancy in 2021.

9.7 Last Words

At mid-2021, the world is still ravaged by the coronavirus, the availability of vaccines is uneven, and borders in many parts of the world remain closed. The world's most privileged economies, including those in China, North America, and western Europe, are rebounding soonest from the pandemic. Due to the interdependencies of the modern world, we know that no one region will be safe from covid-19 until people in all countries have access to vaccines and are mostly vaccinated. Public health experts predicted a pandemic of the SARS-Cov-2 variety, and more pandemics could visit the planet during the twenty-first century. With growing human population, habitats for wildlife shrink and animals become our closer neighbors. This increases the risk of zoonotic transmission of disease.

While we collectively recover from the covid-19 pandemic, the world faces another crisis, this one entirely manmade. Since the industrial revolution, we have collectively emitted so much CO_2 into the atmosphere that a certain amount of global warming is "locked-in" by molecules already released that will remain in the atmosphere for decades at a minimum and centuries for some portion of them. We have known about this problem at least since the early 1990s but have not mustered the political will to address it in a serious way. Climate scientist Jean Jouzel said in 2018 that "in 2020 the objective of limiting warming to less than 2 °C by 2100 compared to the pre-industrial era is no longer realistic; we are on a path at the global level to 3.5–4 °C."

The challenge that faces us today is proving that the collective will exists to decarbonize the world's energy, industrial, and transportation systems, so that by 2050 the world will be "carbon neutral" by virtue of emissions reduced and CO_2 captured and sequestered. Technologies exist today to produce energy with negative carbon intensities. We must bring down their cost and make them widely available.

There is no time to lose, and all must rise to the challenge using the tools of science, technology, and (green) finance to reach our goals.

This book was written by two baby boomers who believe that we all share in the responsibility to achieve a decarbonized future. The early years of the twentieth century in Europe were called the Belle Époque. The period was indeed beautiful for a privileged few whom Thomas Piketty in his book "Capital in the 21st Century" identified as mainly rent-seeking industrialists and their economic enablers (Piketty 2014). At the beginning of the twenty-first century, we need to heed the warnings of ethical leaders like Pope Francis and Mahatma Gandhi. The latter famously said that "Earth provides enough to satisfy every man's needs, but not enough for every man's greed."

Citizens in the most highly industrialized economies bear the most responsibility for the current situation. We must rise to the challenge of the times, especially those in decision-making roles, who often have the tools in their hands but struggle to avert their attention from forces focused on short-term gains, from the lethargy of the greatest number, and the interests of the most powerful. Let the visitation of the pandemic remind us that the world is interconnected and that our fate is inextricably linked to the welfare and success of others.

Jean Hetzel and John C. Shideler

Questions for Readers

1. Whom do you think should account for embodied carbon in products: the producer, the purchaser, or both?
2. Why do critics question the ability of carbon credits to "offset" the emissions of an emitter of greenhouse gases?
3. What role do you think carbon markets should play in the decarbonization of economies?
4. What lessons did pandemic-inspired economic shutdowns have for the fight against climate change?
5. In deciding actions to combat climate change, should lessons and teachings from ethical and spiritual leaders be considered along with perspectives derived from economics, law, engineering, and technology?

References

Ambrose J (2020) Covid-19 crisis will wipe out demand for fossil fuels, says IEA. The Guardian 30 April 2020. https://www.theguardian.com/business/2020/apr/30/covid-19-crisis-demand-fos sil-fuels-iea-renewable-electricity. Accessed 5 June 2020

China Power (2017) How well-off is China's middle class? https://chinapower.csis.org/china-mid dle-class/. Accessed 2 Jan 2021

Davis S, Caldeira K (2010) Consumption-based accounting of CO_2 emissions. https://www.pnas.org/content/107/12/5687/tab-figures-data. Accessed 2 Jan 2021

DNV GL (2015) The state of healthcare. https://issuu.com/dnvgl/docs/dnvgl_-_state_of_healthcare. Accessed 4 Apr 2021

European Commission (2019) The EU's track record on climate action. https://ec.europa.eu/com mission/presscorner/detail/en/fs_19_6720. Accessed 24 Feb 2021

European Commission (2021) A European green deal. https://ec.europa.eu/info/strategy/priorities-2019-2024/european-green-deal_en. Accessed 4 Jan 2021

European Parliament (2018) News. EU progress towards its climate change goals. https://www.europarl.europa.eu/news/en/headlines/society/20180706STO07407/eu-progress-towards-its-cli mate-change-goals-infographic. Accessed 14 June 2021

Fragnière A (2009) La compensation carbone : illusion ou solution? Presses Universitaires de France, also available online at https://journals.openedition.org/developpementdurable/8260. Accessed 2 Feb 2021

Greenpeace (2020) EU 'green' funds should be limited to countries with coal phase outs, 14 January 2020. https://www.greenpeace.org/eu-unit/issues/climate-energy/2558/eu-green-funds-should-be-limited-to-countries-with-coal-phase-outs-greenpeace/. Accessed 14 June 2021

Hausfather Z (2017) Mapped: the world's largest CO_2 importers and exporters. https://www.carbon brief.org/mapped-worlds-largest-co2-importers-exporters. Accessed 4 Jan 2021

Hook L (2020) Greenwash away your sins with these carbon indulgences. In: Financial Times, 2020-12-19, p 11

Howard P, Sylvan D (2021) Gauging economic consensus on climate change, Institute for Policy Integrity New York University School of Law Wilf Hall, 139 MacDougal Street New York, New York 10012. https://policyintegrity.org/publications/detail/gauging-economic-consensus-on-climate-change. Accessed 31 Mar 2021

IPCC (2019) Special report on the ocean and cryosphere in a changing climate. https://www.ipcc.ch/srocc/. Accessed 21 Feb 2021

Marcu A et al (2021) Border carbon adjustment ahead. Proceed with caution. European roundtable on climate change and sustainable transition (ERCST). https://secureservercdn.net/160.153.137.163/z7r.689.myftpupload.com/wpcontent/uploads/2021/03/20210317-CBAM-II_Report-I-Sec tors.pdf. Accessed 8 May 2021

Nguyen T (2020) More companies want to be carbon neutral. What does that mean? In Vox https://www.vox.com/the-goods/2020/3/5/21155020/companies-carbon-neutral-climate-positive. Accessed 2 Apr 2021

OECD (2021) Economic outlook. https://www.oecd.org/economic-outlook/march-2021/. Accessed 5 Apr 2021

Our World in Data (2015). Share of global CO2 emissions vs. share of population https://ourwor ldindata.org/grapher/share-co2-emissions-vs-population. Accessed 7 June 2021

Our World in Data (2020) Per Capita CO_2 Emissions in 2017 vs 1990 https://ourworldindata.org/grapher/per-capita-co2-emissions-1990-vs-2017?tab=chart&stackMode=absolute&country=&region=World. Accessed 4 Feb 2021

Piketty T (2014) Capital in the twenty-first century. The Belknap Press of Harvard University Press, Belknap, Massachusetts

S&P Global (2020a) What is carbon pricing? https://www.spglobal.com/en/research-insights/art icles/what-is-carbon-pricing. Accessed 2 Apr 2021

S&P Global (2020b) Economic research: COVID-19 deals a larger, longer hit to global GDP. https://www.spglobal.com/ratings/en/research/articles/200416-economic-research-covid-19-deals-a-larger-longer-hit-to-global-gdp-11440500. Accessed 15 Feb 2021

Statista (2021) Total value of U.S. trade in goods (export and import) worldwide from 2004 to 2019. https://www.statista.com/statistics/218255/total-value-of-us-trade-in-goods-worldwide-since-2004/. Accessed 23 Feb 2021

Trading Economics (2021) China GDP annual growth rate. https://tradingeconomics.com/china/gdp-growth-annual. Accessed 2 Jan 2021

Chapter 1: Questions for Readers

Question 1–1: What did scientists learn from the Keeling Curve (Fig. 1.4)?

The "Keeling Curve" in Fig. 1.4 records the steady rise in concentration of atmospheric CO_2 since the late 1950s. The graph helps us understand several characteristics of the rise of Northern Hemisphere greenhouse gas emissions over time:

1. The curve trend is steadily increasing on an annual average, with a level of 315 ppm CO_2 in 1968 and a level of 382 ppm CO_2 in 2008. It reached 419 ppm in 2021 at the Mauna Loa Observatory in Hawaii.
2. Emissions are unevenly distributed throughout the year with a peak in May and a minimum in November with a variation of around 5 ppm CO_2.
3. The Mount Pinatubo volcanic eruption shifted the curve on the order of 1 to 2 ppm CO_2.

In summary, the trend shows an increasing concentration of CO_2 in the atmosphere. Natural disturbances can affect the concentration of CO_2 in the atmosphere but do not alter the trend of rising emissions related to human activity.

Question 1–2: What are the key points to understand from data showing average annual CO_2 emission sources and sinks as illustrated in Fig. 1.5?

Figure 1.5 shows the annual fluxes of CO_2 in what is known as the global carbon cycle. Upward pointing arrows represent emissions to the atmosphere. Downward pointing arrows represent carbon sinks in soil, vegetation, and oceans and waterways. This figure shows that fossil fuel development and use add 35 gigatons (Gt—or billion tons) of CO_2 per year to the atmosphere, land-use change adds 6 Gt, and volcanic eruptions 0.5 Gt for a total of 41.5 Gt. Meanwhile, the biosphere absorbs 12 Gt and oceans another 9 Gt. As a result, the stock of CO_2 in the atmosphere increases by 18 Gt/year. Two Gt of anthropogenic emissions remain unaccounted for.

The following table illustrates the carbon cycle imbalance.

J. C. Shideler and J. Hetzel, *Introduction to Climate Change Management*, Springer Climate, https://doi.org/10.1007/978-3-030-87918-1

Period 2009–2018 (source Global carbon project 2020)			
Emissions	Gt/CO$_2$/yr		Removals
Fossil fuel CO_2	35 (33–37)		
Land-use change	6 (3–8)		
		12 (9–14)	Biosphere
		9 (7–11)	Ocean
Increase in atmospheric CO$_2$	**+ 18**		
CO$_2$ unaccounted for	**+ 2**		

Question 1–3: Why are global warming potentials important to understand?

Global warming potentials (GWPs) provide a means for comparing the amount of radiative forcing a greenhouse gas (GHG) will provide compared to carbon dioxide. A GWP is always associated with a time horizon, meaning the amount of radiative forcing a gas will deliver over a given number of years. The most commonly used time horizon—100 years—is specified for countries reporting their national GHG emissions to the UNFCCC. Because GHGs have different residence times in the atmosphere before they chemically degrade, the radiative forcing of a gas may be higher or lower if a different time horizon is selected. The time horizon of 20 years is often used for methane (CH_4) to justify voluntary or regulatory policies associated with oil and gas development or agricultural activities.

Question 1–4: What is the significance of global "hotspots" where warming is occurring at rates higher than the mean?

When scientists and policy makers discuss global warming, the most common approach is to refer to global average temperatures from land and ocean monitoring sites around the world. Thus, targets to limit global warming to 1.5 °C or 2 °C do not preclude warming variations at different locations. Indeed, there are locations around the world where warming has already exceeded the average. We know, for example, that higher than average warming has occurred in Arctic regions and in some locations in the Middle East. Data from September 2020 show surface air temperature anomalies around the world (Fig. 1–10).

Question 1–5: How are the Representative Concentration pathways developed and how should they be interpreted?

RCPs are created using integrated assessment models (IAMs), and they project outcomes up to the year 2100 under differing GHG emission scenarios. Extended concentration pathways (ECPs) extend the RCPs from 2100 to 2500 based on stakeholder consultations and professional estimates. Figure 1.9 tracks the rise of global CO_2 emissions from 1960 to 2018 and projects future emissions according to different RCPs. RCPs provide policy makers with a basis for planning mitigation targets. Risk managers use RCPs to anticipate potential levels of financial losses and to make the necessary provisions to offset them. Investors use RCPs to confirm or inform the potential payback of an investment. RCPs are useful projections, but they do not provide localized information. For example, Arctic Circle regions in Alaska, Canada, and Russia are warming at approximately twice the rate of the "representative" pathway concentrations.

Chapter 2: Questions for readers

Question 2–1: What do you think the effect was of the Kyoto Protocol coming into force without the participation of the USA and China?

The absence of the USA and China as participants in the Kyoto Protocol retarded collective action to combat climate change at the international level. In the period from the signing of the Kyoto Protocol in 1997 to the signing of the Paris Agreement in 2015, China became the world's largest emitter of GHGs. The absence of the USA and China from the Kyoto Protocol prompted Japan to announce that they would not join the Protocol's second commitment period (2013–2020). Canada subsequently withdrew from the treaty. The European Union retained its commitment through the second commitment period. Lack of universal adherence among developed economies to the Kyoto Protocol led to its replacement in 2015 by the Paris Agreement which was approved by 195 countries, including the USA and China.

Question 2–2: What are the inherent limitations of cap-and-trade programs that do not include as participants all industrialized economies?

Cap-and-trade programs that are not universally applied can distort international trade by giving companies located in unregulated economies a cost advantage over companies located in regulated economies. Such cost disparities can encourage owners of industrial facilities in regions subject to cap-and-trade programs to relocate to regions that do not use this mechanism to reduce GHG emissions. The interest countries have to retain their industrial infrastructure has led policy makers to provide free emissions allocations to facilities in trade-impacted sectors. It has also resulted in efforts to impose carbon border adjustment mechanisms to level the economic playing field among countries that participate in cap-and-trade programs and those that do not.

Question 2–3: What relevance do declines in per-capita GHG emissions have on the ability of the world to keep global warming to between 1.5 and 2 °C?

To meet Paris Agreement goals to reduce GHG emissions and stabilize and then reduce global warming, per-capita GHG emissions must decline. However, absolute greenhouse gas emissions are also affected by rising world populations. If per-capita GHG emissions do not fall faster than the rate at which population increases, absolute GHG emissions will not fall—or fall fast enough—to meet Paris Agreement targets for reduced global warming.

Question 2–4: What do you believe the relative advantages and disadvantages are between the imposition of carbon taxes and the establishment of cap-and-trade systems?

Carbon taxes have the advantage of applying universally to the consumption of carbon-intensive raw materials such as coal, petroleum, and natural gas. Carbon taxes, however, disproportionately affect the living standards of low-income persons whose expenditures on fuels for transportation and heating take up a greater share of their disposable income than those with higher incomes. Cap-and-trade systems provide economic incentives to industries to decarbonize. Facilities that can decarbonize faster and to a greater extent than their competitors are rewarded with reduced compliance costs. (These may be offset, however, by their investment costs in decarbonization.) In theory, by setting an overall cap in emissions, policy makers achieve specific emission reductions at the least overall economic cost by allowing market forces to determine which facilities will decarbonize soonest and to what extent. The disadvantage of cap-and-trade systems is that policy makers may set caps at too high a level to achieve meaningful GHG reductions, or they may provide free allowances to facilities deemed to be impacted by international trade from unregulated competitors, thus weakening the effectiveness of the system. The effect of both carbon taxes and cap-and-trade systems should be to encourage the substitution of low-carbon alternatives for fossil fuels and petroleum-based feedstocks.

Question 2–5: What is the importance of sending a "price signal" for emissions of greenhouse gases, and the level of that signal?

Economic choices are based on need and preference. Demand will generally be higher for goods that are lower priced than their alternatives after taking into account quality and overall value. Typically, low-carbon alternatives have higher costs than the goods they are intended to replace. Policy makers, therefore, can send a price signal by taxing the high-carbon alternative or by providing incentives to the producers of low-carbon alternatives. The objective is to reduce the cost disparity between the alternatives and to encourage adoption of low-carbon goods. With price signals sent, investors will allocate money to low-carbon goods, and they will become more widely available. Production of low-carbon goods at greater scale typically leads to reductions in unit costs, making the low-carbon choice more economically competitive.

Chapter 3: Questions for readers

Question 3–1: Why is it important for an organization to set a boundary (organizational, reporting) for its emissions inventory?

A boundary defines the scope for GHG inventory reporting and is an essential element of GHG reporting at the organizational level. Emission sources that fall within the boundary are reported, those that do not, are not. Single facilities may be required by regulation to report at the facility level (e.g., the US EPA's mandatory greenhouse gas reporting regulation in the USA). Reporting boundaries set at the corporate level typically include all emissions for an organization including its domestic and international facilities. In the latter case, emissions from facilities, from corporate headquarters, and from vehicle fleets are all reported to a headquarters manager who consolidates the emissions and reports them on behalf of the organization. It is common in some sectors, such as oil and gas exploration and production and electricity generation, to report fractional shares of emissions at facilities with more than one owner.

Question 3–2: What is the difference between GHG emissions and GHG removals?

Emissions come from direct or indirect sources and are categorized as resulting from combustion; chemical processes; leaks or releases ("fugitive" emissions); and land use, land-use change, and forestry. Removals occur when carbon is sequestered through biological processes such as photosynthesis and by deposition and accumulation in soil. Removals are primarily associated with the agriculture and forestry sectors. Direct capture of CO_2 from the atmosphere is another type of removal. If an organization quantifies GHG removals, it will identify and document the corresponding GHG sinks such as forests, grasslands, and agricultural soil. It should be noted that for inventory reporting only the increases in removals from a prior period should be quantified and not the total stocks of carbon in the sink. From an inventory accounting perspective, emissions and removals do not offset one another. Instead, they are separately quantified and reported within the defined organizational reporting boundary.

Question 3–3: Why is it important to involve a variety of organizational functions and personnel in the quantification of greenhouse gas emissions?

Reporting of organizational GHG emissions and removals can be a complex task. It requires accessing data that may be stored in databases operated by many different departments. Examples include real estate divisions to understand what facilities an organization owns. Vehicle information may be reported to a fleet manager. An operations center may have data about corporate use of aircraft. Facilities managers typically review cost and consumption data for utilities. Asset managers track fixed equipment. Operational personnel maintain measurement devices and operations managers review production data. Shipping and receiving personnel can be a source for transportation data. It often falls on environmental management personnel to report GHG emissions, but these individuals must have access to relevant information from across the organization.

Question 3–4: What are the characteristics of an environmental management system that make its implementation useful to organizations that quantify their greenhouse gas emissions?

A management system is a set of interrelated or interacting elements of an organization to establish policies and objectives and processes to achieve those objectives. An environmental management system may be based on the requirements in ISO 14001. It should identify greenhouse gases as an environmental aspect and use the system to achieve the following goals:

1. Develop an organized, consistent system based on a Plan-Do-Check-Act process that ensures the planning, implementation, control, and continuous improvement of the system. To fulfill planning requirements, organizations should identify GHG sources and sinks, quantify emissions and removals, and set objectives for their respective reduction or enhancement.
2. Continually improve the organization's environmental performance, including by reducing emissions and increasing removals according to the policy and objectives of the organization.

Question 3–5: What is the value to consumers of requiring disclosures of product environmental footprint information on products sold at retail?

Information empowers consumers to make purchasing choices based on their environmental preferences. For example, consumers may purchase paper made with recycled content as a way of supporting a circular economy and reducing the amount of trees harvested to provide pulp for paper making. Displaying a product's carbon footprint can lead environmentally conscious consumers to select lower carbon intensity products. For product comparisons to be valid, the products must be functionally equivalent. The carbon footprint must be established based on the same life cycle assessment rules and account in identical ways for co-products, waste, and end-of-life fates.

Chapter 4: Questions for readers

Question 4–1: Why is the concept of additionality easier to understand than apply?

The concept of additionality seeks to prevent emission reductions or removal enhancements from being granted carbon credits if the activities that produced them would have occurred anyway, without the financial incentives that carbon credits provide. The difficulty is defining the threshold at which the activity swings from being worthy of economic support to being "business as usual." When assessing whether a particular activity should be considered "additional," carbon credit programs typically assess such variables as common practice, regulatory requirements, and technical or economic barriers. Determinations of additionality are inherently subjective, and for this reason are often debated.

Question 4–2: Why do investors and others insist that mitigation projects be independently validated and verified?

Mitigation projects that result in the issuance of carbon credits only have value to investors if they believe that the project activities have reduced emissions or enhanced removals in a scientifically demonstrable way. In other words, they seek assurance that the metric ton of GHG emission reduction or removal enhancement is real, permanent, additional, quantifiable, verifiable, and enforceable. Validation provides assurance that the project activity is capable of reducing emissions or enhancing removals, and that the activity should be recognized as a legitimate mitigation project. Verification assesses historical data from project activity and substantiates the existence of the project and its accurate quantification of emission reductions and removal enhancements during a specified reporting period. Carbon offset registries rely upon verification reports to make final decisions on the issuance of carbon credits.

Question 4–3: Why is a ton of CO_2 removed from the atmosphere equivalent to a ton of CO_2 emission reductions?

Global warming is the result of an increased amount of greenhouse gases residing in Earth's upper atmosphere. The global carbon cycle includes emissions into the atmosphere and removals from the atmosphere. Each is measured in carbon dioxide-equivalent metric tons. Reducing emissions and enhancing removals both have the effect of bringing the global carbon cycle closer to balance.

Question 4–4: What is the difference between a carbon credit issued under regulatory and voluntary GHG programs?

Carbon credits issued under regulatory cap-and-trade programs or low-carbon fuel standards are similar to those issued under voluntary programs. The main difference is in their application. Regulatory carbon credits are used to meet compliance obligations under a cap-and-trade or *bonus-malus* (e.g., credit-deficit) program. The other difference is price. Regulatory carbon credits used in compliance programs generally trade at higher monetary values than voluntary carbon credits. In either case, emission reductions and removal enhancements should be real, permanent, additional, quantifiable, verifiable, and enforceable and denominated in metric tons of CO_2-equivalent.

Question 4–5: How do GHG programs take into consideration environmental objectives other than GHG emissions and removals?

Clean Development Methodologies since 2003 have included provisions that Project Design Documents describe the results of the environmental impact assessment of projects when required "by the host Party [country]." In most cases, formal environmental impact assessments have not been required as project designs have appeared to provide environmental benefits with few or no negative environmental impacts. The CDM process also includes a provision for stakeholder consultation. This mechanism may or may not result in actionable modifications to the project design. The Gold Standard and the Climate, Community, and Biodiversity standard in the voluntary market have made conscious efforts to require and document co-benefits besides greenhouse gas mitigation. The benefits may be social, such as providing employment to local residents, or environmental, such as enhancing biodiversity. The Verified Carbon Standard includes a "Safeguards" clause that states that "Project activities shall not negatively impact the natural environment or local communities." The VCS also requires local stakeholder consultation prior to project validation and a mechanism for ongoing communication with local stakeholders during project implementation.

Chapter 5: Questions for readers

Question 5–1: Why is the use of activity data and emissions factors the dominant method for quantifying GHG emissions?

Activity data describe substances or processes that will generate GHG emissions. Liters or cubic meters of fossil fuels are a good example. It is easy to add up the total volume of fossil fuels an organization purchases in a year. But these numbers do not provide us with equivalent tons of carbon dioxide (CO_2) emissions. For that, an inventory manager needs emission factors. Emission factors convert a stated amount of combusted fuel into emissions of CO_2, methane (CH_4), and nitrous oxide (N_2O). Global warming potentials (GWPs) then normalize the emissions of methane and nitrous oxide into "carbon dioxide-equivalent" (CO_2e) emissions. Emission factors for the consumption of petroleum-based fuels are fairly reliable as they are based on stoichiometric analyses of the combustion process. Emission factors for other fuels may be less reliable, such as for wood combustion, where the amount of moisture in the wood is an important variable. In any case, most GHG inventories rely on activity data and emission factors for the quantification of GHG emissions as these methods are easier to execute and obtaining information from direct measurement is often not feasible or cost-effective.

Question 5–2: How important is calibration and maintenance of measurement and monitoring devices in ensuring accurate quantification of GHG emissions?

When activity data are directly measured, such as by a flow meter or gas analyzer, calibration ensures that reported values are within reasonable tolerances for accuracy. The manufacturers of flow meters usually calibrate them within a tolerance of $\pm 1\%$

- 2%. Similarly, gas analyzers can be calibrated within a manufacturer's tolerance, but an added uncertainty is the accuracy of a calibration gas which is an independent variable that may either reduce or exacerbate any detected "drift" of a gas analyzer from its factory calibration. Generally speaking, the importance of calibration accuracy and the tolerance for calibration drift is a function of the purpose for quantifying greenhouse gas emissions. Regulatory programs or carbon offset reporting programs can require overall measurement accuracy to be within 1% to 5% of the true value.

Question 5–3: In what ways does validation differ from verification?

Validation, according to ISO 14064 Part 3, Greenhouse gases—Specification with guidance for the verification and validation of greenhouse gas statements, is a "process to evaluate the reasonableness of the assumptions, limitations, and methods that support a statement about the outcome of future activities." It occurs either before a mitigation project is implemented, or shortly afterward. The purpose of validation is to provide an opinion, based on a project design document, that the project as planned is likely to produce the intended emission reductions or removal enhancements, and is likely to meet the applicability requirements contained in the selected methodology. Verification is backward looking and occurs at selected intervals after project validation. A GHG program or the project developer defines reporting periods when the project has monitored and quantified emission reductions or removal enhancements. The information being verified is therefore historical in nature. Verifiers determine that emission reductions and removal enhancements occurred based on evidence gathered and reviewed. Verifiers also check to make sure that the information provided by the person responsible for quantifying the emissions reductions or removal enhancements (the "responsible party") was complete and accurate, properly classified, and correctly assigned to the stated reporting period.

Question 5–4: In what ways does a limited assurance verification engagement differ from a reasonable assurance verification?

Limited assurance verification engagements provide intended users of verification opinions with a lower level of confidence in the accuracy of stated emissions or removals. While the basic steps in the verification process are the same—strategic analysis, risk assessment, planning, execution, and completion—the focus of a limited level of assurance engagement is on consolidated statements of emissions and removals rather than on emissions and removals at the level of individual sources and sinks. Site visits may be omitted in limited level of assurance engagements, particularly if the verification team already has knowledge of the responsible party's facilities and its data and information systems. When site visits occur, they are generally made at the location where organizationwide emissions are reported and consolidated.

Question 5–5: How do agreed-upon procedures differ from verification?

A new feature of ISO 14064–3:2019, *Greenhouse gases—Specification and guidance for the verification and validation of greenhouse gas statements,* is Agreed-upon procedures (AUP). Annex C describes how verifiers may use the activities and techniques of verification when intended users do not require an opinion. It is for specialized applications only and is not a substitute for an assurance engagement. The relationship of AUP to engagements that provide assurance through the issuance of a verification opinion, and validation engagements that express an opinion, is clearly differentiated.

According to ISO 14064–3:2019 Annex C.2, AUP are useful in the following circumstances:

a) GHG programs that specify agreed-upon procedures rather than assurance
b) specific indirect emissions and removals (indirect emissions in inventories; upstream and downstream emissions and removals for product life cycles)
c) compliance to specifications.
d) greenhouse gas information and data management and controls.

Chapter 6: Questions for readers:

Question 6–1: Is a blockchain cryptocurrency a money?

Cryptocurrencies are a form of money, even if they are not legal tender. The U.S. Federal Reserve System includes on each note the phrase "This note is legal tender for all debts, public and private." Historically, the right to issue money has been reserved to sovereigns, either monarchs or national governments. The value of money issued by sovereigns is based on the trust that people have in its issuer and its stability. The value of cryptocurrencies relies on perceptions of its scarcity compared to purchasers' desires to hold it.

Question 6–2: Is the current ratio of investment from public and private sources scalable to the economywide decarbonization needs of the years up to 2050?

According to the Climate Policy Initiative, the average expenditures of public sector funds on climate finance in 2017 were approximately \$275 billion. This figure included funds dispensed by multilateral banks whose seed capital was initially obtained from governments but whose actual funds available for grants and loans have been increased by profits made from lending activities. If the total need for investment in climate change mitigation and adaptation from 2016 through 2030 was approximately \$90 trillion (as estimated by the OECD), the annual amount of needed funding would be \$6 trillion per year. The \$275 billion of public money expended in 2017–2018 represents one half of one-tenth of a percent (0.05%) of this \$6 trillion. It is unlikely that public sector would have the resources to fill such an investment gap. This highlights the importance of harnessing private-sector financing for climate change mitigation and adaptation needs.

Question 6–3: Do all types of actors in the finance sector have equal roles to play in greening world economies?

It is hard to imagine how mitigation and adaptation needs can be met without very broad participation of economic actors throughout the world and in every economic sector. Developed countries have more resources available to them than developing countries do, so it seems logical that they will play a larger role on a per-capita basis in financing the transition to a low-carbon economy. Nonetheless, investible wealth exists in all economies, regardless of development status. Successful climate change mitigation and adaptation efforts will need access to all sources of capital wherever located during the transition to a low-carbon economy.

Question 6–4: How does the ISO 14030–3 taxonomy take environmental aspects into account?

The ISO 14030–3 taxonomy requires a "do no significant harm" analysis that prompts the consideration of environmental aspects when selecting projects, assets, and supporting expenditures for investment. The taxonomy facilitates decision making on sustainable economic activities. Projects, assets, and supporting expenditures described in the taxonomy consider relevant environmental objectives and trigger consideration of environmental aspects and issues.

Question 6–5: What should the role be for green finance in the decades leading up to 2050?

Green finance is a critically important enabler for the transition to a low-carbon or decarbonized economy. Every sector of the economy that emits greenhouse gas emissions directly, through its supply chain, or via the use of its products and services faces decarbonization challenges. Low-carbon solutions must be found to achieve "net-zero" emissions by 2050. In nearly every case, these solutions will require a degree of financing to be implemented.

Chapter 7: Questions for readers

Question 7–1: How are climate change mitigation and adaptation related?

Climate change that results from anthropogenic (human-induced) emissions of greenhouse gases is impacting both natural and built environments. The scale and frequency of impacts are greater than historical patterns of impacts from acute weather events, biodiversity loss, disease vectors, migration patterns, and economic dislocations. To the extent that greenhouse gas emissions are reduced, adaptation needs may be met more readily. However, even achieving "net-zero" emissions by 2050 will not be enough to avoid additional global warming, so adaptation efforts will need to continue and meet increasingly strict resilience criteria.

Question 7–2: Can you identify situations in which liquefaction has impacted natural landscape features or the built environment?

Liquefaction can attack the soil-supporting buildings or other structures leading to the collapse of structures and the washing away of soils. Intense precipitation events make liquefaction of soils more likely to occur. One of most dramatic examples of liquefaction occurred in Germany in summer 2021 when an area of low atmospheric pressure stalled over western Germany and eastern Belgium causing unprecedented levels of precipitation. As much as a month's average rainfall fell within 24 h, causing rampant flooding that swept away buildings and roads and caused the deaths of approximately 200 persons. The melting of glaciers in the Indian Himalayas in February 2021 was blamed for catastrophic flooding in a tributary of the Ganges River that liquefied riverbanks and destroyed a hydroelectric dam that was under construction.

Question 7–3: To what extent do you believe that your personal residence is adapted to withstand extreme weather events?

The answer to this question depends on many regional and site-specific considerations. Usually, single-family and multifamily homes are built to withstand a locally appropriate range of risks and weather conditions. However, many local building codes have been developed using statistics about historical weather events in the affected community and region, and climate change is making forecasts based on historical conditions inapplicable for current risk levels and those 10 to 30 years in the future. Buildings erected near coastlines will be more exposed in the future to rising sea levels. Precipitation events including tropical storms are predicted to become more intense as ocean and surface temperatures rise and clouds hold more rain. Rivers and streams will expose structures built in floodplains and adjacent to floodplains to more frequent and severe flooding than in the past. A risk analysis developed in conjunction with local adaptation planning is recommended when siting new structures or investing in those previously built.

Question 7–4: How do you assess the trade-offs of living near water?

The answer to this question must be based on an individual's tolerance for risk, as well as the specific details of properties under consideration. In southeastern France in October 2020, severe flooding damaged or destroyed structures that planning maps located outside the floodplain of the Vallée de la Vésubie (Department of Alpes-Maritimes). This is the lesson of climate change that past experience is not an adequate predictor of future conditions. Shorelines are being eroded by ocean currents and rising sea levels, so structures that were safely built near but not on the ocean front may today be dangerously close. Perhaps a houseboat would provide the benefits of living on or near the water without some of the risk associated with soil conditions, subsidence, and rising sea or inland water levels.

Question 7–5: How would you rate the adaptive capacity of your community or your employer?

Use the five-step exercise below to evaluate the situation in your community or that of your employer. The following five steps are adapted from Annex B of ISO 14090 (2019), *Adaptation to climate change—Principles, requirements and guidelines*:

- characterize the system (use a "systems thinking" approach)
- research possible changes to the climate in your community, based on a risk assessment
- identify thresholds for abnormal events that require adaptation planning
- assess resilience of both the built environment and the organizational and human resources (emergency responders, healthcare providers, utilities, etc.)
- identify suitable indicators for judging that infrastructure elements are sufficiently adapted

Implement the resulting hardening strategies and plans.

Chapter 8: Questions for readers

Question 8–1: What role should there be for nuclear energy in the low-carbon economy?

The answer to this question will depend upon the value trade-offs of whom you ask. Some people view nuclear energy as a low-carbon and reliable base-load energy source and for those reasons advocate its expanded use. Others place a higher importance on nuclear non-proliferation, waste management, and safety concerns, and believe its role in the energy mix should be reduced and eventually eliminated. Others view nuclear power generation as uneconomic energy source since the cost of construction of new nuclear facilities cannot be financed without generous government subsidies and favorable permitting regimes. No matter what your personal views are, it is highly unlikely that nuclear energy will play a significant role in the energy future of democratic societies without an increased social consensus in its favor and the willingness of governments to help finance its deployment.

Question 8–2: How should the housing and commercial building sectors be re-imagined to ensure that newly constructed buildings meet high energy-efficiency standards and climate resilience?

Housing and commercial buildings need to do more to take climate change into account. They need to contribute to climate change mitigation by utilizing "zero-emissions" technologies. They also need to prepare for future climate challenges by meeting stricter requirements for resistance to extreme weather. Architects and engineers should question the adequacy of building codes that are based only on backward-looking statistics of wind, weather, and flooding. What history leads us to define as a "100-year flood" or a "500-year flood" is becoming increasingly common. Climate change will continue exacerbating such trends and the built environment will face increasingly severe impacts.

Question 8–3: What role should green finance play in decarbonizing world economies?

Labeling finance as "green" brings purpose and transparency to investment decisions. "Conventional" financing can accomplish the same purpose as labeled green finance. However, investors increasingly recognize the importance of mitigating climate change and increasing resilience through adaptation strategies and are looking for investible projects and assets. The need is enormous. Worldwide investments of several trillion US dollars per year are needed to meet Paris Agreement goals by 2050. Success in transitioning to a low-carbon economy without contributions of capital held by private-sector investors is unimaginable.

Question 8–4: What challenges confront manufacturers in transitioning to low-carbon products?

The greatest challenge facing manufacturers is economic. Low-carbon alternatives exist in nearly every sector of economic activity. Climate-friendly products usually cost more, at least at first until economies of scale can be realized, retooling of manufacturing plants are amortized, and low-carbon product demand rises either through the imposition of carbon taxes on legacy alternatives or until consumer values and behavior change. In some particularly carbon-intensive industries, such as the manufacture of steel and cement, technological challenges remain to be solved. Many solutions to high-carbon products rely on the availability of low-carbon electricity. There is a sufficiently high number of economic and technological variables that coordination of the transition through the introduction of government policies is likely to be needed.

Question 8–5: Can continuing business-as-usual lead the world to a net-zero future?

No.

Chapter 9: Questions for readers

Question 9–1: Whom do you think should account for embodied carbon in products: the producer, the purchaser, or both?

Producers should provide information about the carbon footprint of their products. Accountability does not stop there, however. The purchasers of products, whether they be manufacturing intermediaries or end users, should understand the contribution that embodied carbon makes and seek low-carbon options. Life cycle assessment and carbon footprint of product information should inform purchasing choices, along with suitability for use and cost. Exporting carbon emissions to third country manufacturing sites does not resolve the climate crisis. Emissions need to be accounted for no matter where their occur in the supply chain or in the use phase of products.

Question 9–2: Why do critics question the ability of carbon credits to "offset" the emissions of an emitter of greenhouse gases?

The twenty-first century challenge for inhabitants of planet Earth is to make absolute reductions in greenhouse gas emissions, on a worldwide basis, with the goal of achieving net-zero emissions by 2050. This requires an increased level of accounting for greenhouse gas emissions and commitments to their reduction. With growing populations, the needs of people around the world can only be met on a sustainable basis if societies decarbonize every sector of economic activity. With fragmented accounting of greenhouse gas emissions, a reduction here or there of tons of CO_2-equivalent emissions may well satisfy criteria of "real, permanent, additional, quantifiable, verifiable, and enforceable" and still not contribute meaningfully to solving the climate crisis. Such reductions are important, but they are not at present a sufficient indicator of progress toward achieving the net-zero goal. Our accounting methods need to evolve to measure progress on decarbonization of economic sectors, not simply on reduced GHG emissions.

Question 9–3: What role do you think carbon markets should play in the decarbonization of economies?

Verified offset credits issued by recognized regulatory and voluntary programs can represent real, permanent, additional, quantifiable, verifiable, and enforceable emission reductions. But this market mechanism by itself is not sufficient. In the first two decades of the twenty-first century, the price of offset credits was too low to properly incentivize industry to decarbonize its own processes and products. The additionality of some project offset credits has been legitimately questioned. Regulatory cap-and-trade programs have not covered all sectors and emissions, including those associated with imported goods that cross borders without a carbon tax. Market mechanisms have a role to play in the transition to a low-carbon economy, but they are not the only policy option, and should be complemented with other instruments.

Question 9–4: What lessons did pandemic-inspired economic shutdowns have for the fight against climate change?

The SARS-Cov2 pandemic demonstrated that large declines in greenhouse gas emissions could result from significant changes in the behaviors of people. But how change occurs also matters. The pandemic caused large drops of employment and decimated certain economic sectors such as travel, leisure, and hospitality. Office building occupancy fell. Gross domestic product declined in most countries, though unevenly by sector. The challenge for the recovery phase of the pandemic is to promote economic activity that contributes to growth while reducing greenhouse gas emissions on an absolute as well as on a per-capita basis. Society needs social safety nets for reasons of equity and stability. The pandemic also demonstrated how research and development could produce vaccines against SARS-Cov2 in record time. The world needs to marshal a similar commitment and dedication to addressing the climate crisis.

Question 9–5: In deciding actions to combat climate change, should lessons and teachings from ethical and spiritual leaders be considered along with perspectives derived from economics, law, engineering, and technology?

It is fair to say that public policy in general tends to reflect values from economics, law, engineering, and technology. A not insignificant proportion of society, especially in developed countries, have relatively high confidence in the ability of technology to address the world's most pressing problems. The climate crisis has sometimes been framed as one of generational equity, where post-World War II populations created societies economically supercharged by fossil fuel exploitation and left to generations coming of age in the twenty-first century the responsibility to take care of the ensuing environmental crisis. There is merit in this argument, though it focuses more on assigning responsibility for past actions than on crafting forward-looking solutions. It is unlikely, however, that technology can provide all the solutions needed to combat climate change. In our view, social equity considerations need to inform and guide the transition to a low-carbon economy. Individual behaviors and values also need to change, and for this reason, persons who speak from positions of moral authority are critically important.

Index

Printed in the United States
by Baker & Taylor Publisher Services